INVESTING IN
FUTURES & OPTIONS MARKETS

Investing in Futures & Options Markets

Lowell B. Catlett, Ph.D.
Professor of Agricultural Economics and Agricultural Business
College of Agriculture and Home Economics
New Mexico State University

James D. Libbin, Ph.D.
Professor of Agricultural Economics and Agricultural Business
College of Agriculture and Home Economics
New Mexico State University

an International Thomson Publishing company I(T)P®

Albany • Bonn • Boston • Cincinnati • Detroit • London • Madrid
Melbourne • Mexico City • New York • Pacific Grove • Paris • San Francisco
Singapore • Tokyo • Toronto • Washington

NOTICE TO THE READER

Publisher does not warrant or guarantee any of the products described herein or perform any independent analysis in connection with any of the product information contained herein. Publisher does not assume, and expressly disclaims, any obligation to obtain and include information other than that provided to it by the manufacturer.

The reader is expressly warned to consider and adopt all safety precautions that might be indicated by the activities described herein and to avoid all potential hazards. By following the instructions contained herein, the reader willingly assumes all risks in connection with such instructions.

The publisher makes no representations or warranties of any kind, including, but not limited to, the warranties of fitness for particular purpose or merchantability, nor are any such representations implied with respect to the material set forth herein, and the publisher takes no responsibility with respect to such material. The publisher shall not be liable for any special, consequential, or exemplary damages resulting, in whole or in part, from the readers' use of, or reliance upon, this material.

Cover Design: Carolyn Miller

Cover images courtesy of PhotoDisc

Delmar Staff:
Publisher: Susan Simpfenderfer
Acquisitions Editor: Jeff Burnham
Developmental Editor: Andrea Edwards Myers
Production Manager: Wendy A. Troeger
Production Editor: Carolyn Miller
Marketing Manager: Katherine Hans

Printed in the United States of America

For more information, contact:

Delmar Publishers
3 Columbia Circle, Box 15015
Albany, New York 12212-5015

International Thomson Publishing Europe
Berkshire House
168–173 High Holborn
London, WC1V7AA
United Kingdom

Nelson ITP, Australia
102 Dodds Street
South Melbourne
Victoria, 3205 Australia

Nelson Canada
1120 Birchmount Road
Scarborough, Ontario
M1K 5G4, Canada

International Thomson Publishing France
Tour Maine-Montparnasse
33 Avenue du Maine
75755 Paris Cedex 15, France

International Thomson Editores
Seneca 53
Colonia Polanco
11560 Mexico D. F. Mexico

International Thomson Publishing GmbH
Königswinterer Straße 418
53227 Bonn
Germany

International Thomson Publishing Asia
60 Albert Street
#15-01 Albert Complex
Singapore 189969

International Thomson Publishing Japan
Hirakawa-cho Kyowa Building, 3F
2-2-1 Hirakawa-cho, Chiyoda-ku
Tokyo 102, Japan

ITE Spain/Paraninfo
Calle Magallanes, 25
28015-Madrid, Spain

1 2 3 4 5 6 7 8 9 10 XXX 04 03 02 01 00 99

Library of Congress Cataloging-in-Publication Data
Catlett, Lowell B.
Investing in futures & options markets / Lowell B. Catlett, James D. Libbin.
p. cm.
Includes bibliographical references and index.
ISBN 0-8273-8570-6
1. Futures. 2. Options (Finance) I. Libbin, James D. II. Title.
HG6024.A3C38 1999 98-33925
332.63′228—dc21 CIP

For
Joni Marie
and
Mary

Contents

List of Figures

List of Tables

Preface

No one foresaw today's futures and options markets quite as well as the founders of today's Chicago Mercantile Exchange. In what proved to be a classic understatement, the Butter and Egg Board changed to the Chicago Mercantile Exchange in 1919 because they felt that one day they would trade things other than butter and eggs! Yet even those early visionaries would be astounded by what the futures and options markets can and are doing for millions of investors the world over.

The early use of futures contracts in the United States was a direct result of changes in the grain system. The cash grain market needed smoother flows of grain to market, and time contracts started to fill the void. Gradually the industry found that to attract capital and more liquidity, more-sophisticated contracts would be needed. And just as importantly, more-sophisticated use of the existing contracts would be needed also. This pattern has repeated itself from grains to livestock to currencies to stock indexes.

As more and more investors are attracted to the futures and options markets, the uses of the contracts become more complicated and the strategies more sophisticated. However, investment strategies in futures and options contracts do not have to be either complicated or sophisticated to be powerful. All that is required by investors is a fundamental basic knowledge of a few powerful concepts. In fact, some of the recent well-publicized failures of organizations and companies that used futures and options contracts were the result of not knowing the underlying concepts.

Throughout this book we emphasize the fundamental basics of investing with futures and options contracts by providing not only the underlying concepts but enough examples to see how the process works. We purposefully cover most types of futures and options contracts. The world of global business means that a grain company is not immune to the risks of currencies or financial risks. Increasingly, businesses and individuals must view the world of investment with a broader look than a single industry. When the underlying concepts are well known and used, then more-sophisticated uses will naturally develop because more investment risks will be identified.

We gratefully acknowledge the help and input of many students over the years in the development of this book. Special thanks goes to Irma Marshall. Her professional talents moved the book from rough ideas to a finished product with speed and skill. The authors and Delmar Publishers also thank the numerous reviewers and colleagues who provided valuable insights on ways to improve the book, and we sincerely appreciate their valuable time and advice.

Colin Carter
University of California–Davis
Davis, CA

Darren Frechette
The Pennsylvania State University
University Park, PA

Robert Herrmann
The Pennsylvania State University
University Park, PA

Jimmy T. LaBaume
Sul Ross State University
Alpine, TX

David B. Narrie
Tennessee Technological University
Cookeville, TN

CHAPTER 1

Getting Started

"People are always blaming their circumstances for what they are. I don't believe in circumstances. The people who get on in this world are the people who get up and look for the circumstances they want and if they can't find them—make them!"

—George Bernard Shaw

Introduction

During the last decade, investor interest in futures markets, especially as indicated by the volume of futures trading, has exploded. The bread and butter of U.S. futures markets used to literally be bread and butter—grains, livestock, and soybeans. However, in addition to the traditional agricultural commodities, there are very popular new futures contracts on foreign currency, debt instruments, stock indexes, petroleum products, and metals. A companion market in options on futures contracts has existed since the early 1980s. Likewise since the late 1980s contracts that derive their value from futures and options, called derivatives, have become very popular.

The growth in both futures and options contracts has been nothing short of spectacular for some very good reasons. In the early 1970s, the USSR entered the U.S. grain market, droughts occurred across many parts of the United States, transportation bottlenecks surfaced, and a whole host of other things happened, all more or less simultaneously, to cause prices of major agricultural commodities to fluctuate more within the year than they had during the previous twenty years. Trading volume in the futures markets increased accordingly. Prior to the turmoil of the 1970s, it was unusual for the cash or futures of corn to fluctuate more than 1¢ per bushel per day; but now it is not unusual for the fluctuation to be as much as 20¢ per bushel per day.

The severe 1970s and 1980s produced other dramatic changes in the general economy that would later shake up futures markets. Federal restrictions on gold prices and gold ownership were removed, the United States abandoned the Bretton-Woods Agreement and let the dollar float on international monetary exchanges (or sink, as some would say), inflation and interest rates reached double digits, and international trade entered an era of increased activity and change. The price stability of the 1950s and 1960s was gone, and in its place came volatility and uncertainty. The uncertain economic climate during the last two decades has caused futures trading volume to explode. In 1973 the volume of all futures contracts traded in the United States was 26 million contracts, but by 1983 it was 140 million contracts. In 1997 it had soared to 400 million.

Price uncertainty and increased volatility caused investors to speculate on the changing price situation, always looking for another dollar of profit. Also, investors who held the actual commodity or instrument, or who anticipated buying the commodity, needed protection from the changing price situation. They responded by increasing their usage of futures markets by hedging, or shifting the risk of price changes to other traders. Thus, the futures markets offer two major investment opportunities: speculating and hedging.

Investors wanting to speculate on price movements of commodities, stock indexes, debt instruments, and foreign currencies can purchase futures contracts rather than purchase the actual commodities. The main reason an investor in actual commodities (often shortened to the words actuals or physicals) would consider using a futures contract is leverage. Futures contracts are leverage contracts. That is, only a fraction of the contract value must be invested. An investor who believed that gold prices were going to increase, and wished to speculate on that belief, could buy 100 ounces of gold (bullion or coins) and put up $400 per ounce, for a total investment of $40,000. He must pay the entire $40,000. If the price increases in three months to $410 per ounce and he sells, then he has made $10 per ounce, or $1,000 on a $40,000 investment for three months (a 2.5% return [$1,000 ÷ $40,000] for three months or approximated an annualized 10 percent rate of return) less commissions. Control of 100 ounces of gold with a futures contract would require a margin deposit of only approximately $4,000. If the same price increase occurred, the speculator would still capture the $10 increase in price. Thus, the $1,000 return on investment changes from an approximate annualized 10 percent to a 100 percent rate of return, less commissions. Put another way, the $40,000 investment necessary to control 100 ounces of physical gold could be leveraged to control 1,000 ounces of gold through futures contracts.

Leverage, of course, is a two-edged sword. Price movements against the investor's position produce equal but opposite results compared to movements in his favor. If the speculator had invested $40,000 in 1,000 ounces of futures gold, a $10 price decrease would produce a loss of $10,000, or a 25 percent rate of return for the three months the futures position was maintained (a 100 percent loss on an annualized basis). Despite the potential for quick and large losses, leverage is the major attraction of speculative investment in futures contracts.

Leverage is also used by the other major type of investor in futures contracts, the hedger, but leverage is not the major attraction of hedging. Hedgers invest in futures contracts as a method to shift or lay off the majority of the risk of price changes. Although hedgers cannot shift all the risk, they can shift most of it. Thus, there remains a (diminished) element of speculative investing in hedging called basis risk, which is discussed in Chapter 6.

All forms of investment require serious analysis of goals, attitudes toward price risk, and financial resources. In this respect, investing in futures contracts is no different than investing in other traditional assets. To be a successful investor in futures and options, a serious analysis of goals, risk-taking preferences, and resources is necessary, regardless of whether you are interested in being a speculator or a hedger.

Self-Appraisal

As an individual investor, should you invest in stocks? bonds? real estate? diamonds? coins? stamps? Treasury bills? franchises? futures? options? The list of potential forms of financial investment is long and diverse. Each investment has unique advantages and disadvantages, but all investment forms have several common characteristics. To be successful in any investment requires an in-depth self-appraisal of the individual or entity regarding goals, attitudes toward risk, and financial resources.

Setting Goals

The first major question for any investor to ask of herself is: Am I going to be a speculator (someone who trades futures only for a profit) or a hedger (someone who trades futures to protect the price of a cash position)? Although this question sounds trite, it may be the most important. Bankruptcy courts are filled with examples of large firms and individuals that lost sight of this simple question. Large grain companies, cotton dealers, portfolio managers, and individuals have been victims of goal switching. Many start out hedging and gradually begin to speculate. Although there is nothing implicitly wrong with goal switching, the fault lies in not explicitly changing the goal and the corresponding objectives to achieve that goal. Deciding whether to hedge or speculate is the easiest goal to set, and surprisingly is also the easiest to violate. DO NOT TAKE THIS GOAL LIGHTLY. In fact, keep a written record of your goals and review them regularly to remind yourself of your original goal. Update and modify those records as your goals change, but before you change any investment strategy, refer to the historical written goals record. Perhaps an old but useful rule of thumb would be helpful: Individuals will accomplish about 25% of their unwritten goals; however, over 75% of written goals are completed. In other words, write your goal(s) down so that you explicitly know what you want to achieve.

The decision to speculate or to hedge must be fit into the overall goals of the individual or firm. Your overall goals as an individual or firm will include assessments of your attitudes toward risk and your financial resources. Just remember that any unpriced commodity being produced or being stored is actually in a speculative position in the cash market.

Attitude towards Risk

Are you a person or firm that likes to take price risks (are you a risk lover) or do you like the fun and oftentimes higher rewards associated with high-risk investments? Do you sometimes, under certain circumstances, but seldomly, take price risks (are you risk neutral)? Or are you a person who likes to avoid virtually all price risks (are you risk averse)? It is important to analyze yourself or firm and make this decision—it will affect your goals and objectives.

You may love to take price risks in certain situations, such as at the blackjack table or even in business transactions, but do you understand yourself well enough to discipline your trades in the futures market or in other investments? Do you want to take price risks on a portion of your investment and protect the rest? protect it all? Futures and options markets can do both, but only if you have a set of goals, objectives, and risk attitudes defined. Most people are generally risk averse, but each person can be a risk lover under certain circumstances—usually concerning our financial resources.

Financial Resources

If you just inherited $1,000,000, you are more likely to put $50,000 in a risky venture than if your cash flow is tight and you have to scrape to put together the

$50,000 investment. If you are just getting started in a business and your capital is strained, cash flow is tight, and you have very little margin of error, a major price move against you could be disastrous. Hedging may be necessary because of your financial resources. On the other hand, a business that is relatively cash rich has the ability to withstand major price moves without jeopardizing the financial base of the firm.

Developing a Plan

Goals, attitudes toward risk, and financial resources can all be in conflict. One may not be logically consistent with the others; in fact, they usually never are. That is why they must be combined in a plan.

Let's first investigate the case of speculators, the major investors in futures markets. Your goal may be to be a speculator because you love to take price risks, however you don't have the financial resources to invest. Let's say that you are interested in investing in gold. To buy 100 ounces of gold at $400 per ounce requires a $40,000 investment in the actuals, but because you don't have $40,000, you could reduce that investment to $4,000 by switching to a futures contract. Now, do you have a plan? That is, are all three aspects of a plan in agreement? Maybe not. Yes, your goal of speculation is met, you want to take the risk of gold price changes, indeed you are excited about it, and you have $4,000 to get control of a futures contract, but have you met your overall goal? Does this particular plan fit into your total investment goal? If the $4,000 is your life savings or your rent and egg money (a popular expression in the futures trade to denote the money necessary to provide basic living needs), you may want to reconsider the investment.

A helpful way to evaluate the investment is to construct a financial pyramid (Figure 1.1). Place near the bottom of the pyramid the least risky investments that also have high liquidity. Then move up toward the peak of the pyramid with investments that inherently possess more risk and less liquidity. The purpose of the pyramid is to indicate to a risk-averse investor that he should allocate a smaller portion of the total investment portfolio to the more risky and less liquid investments.

The base provides the rent and egg money and safety margin for emergencies. That base might include such investments as passbook savings accounts, savings bonds, Treasury bills, or short-term certificates of deposit. The next layer allows for growth and increases in a relatively safe framework that can be converted to cash, if necessary, fairly quickly and without too much loss in face value, if any. Examples of this intermediate level might include corporate stocks or individual retirement accounts. The smaller, highest portion is for the risky or very non-liquid investments and may include illiquid assets such as mortgages, real estate, or long-term bonds, and risky assets such as gold, diamonds, speculative real estate, futures contracts, and options. The risky parts of this top portion of the pyramid are often called "investments you can afford to lose"—that is, your daily

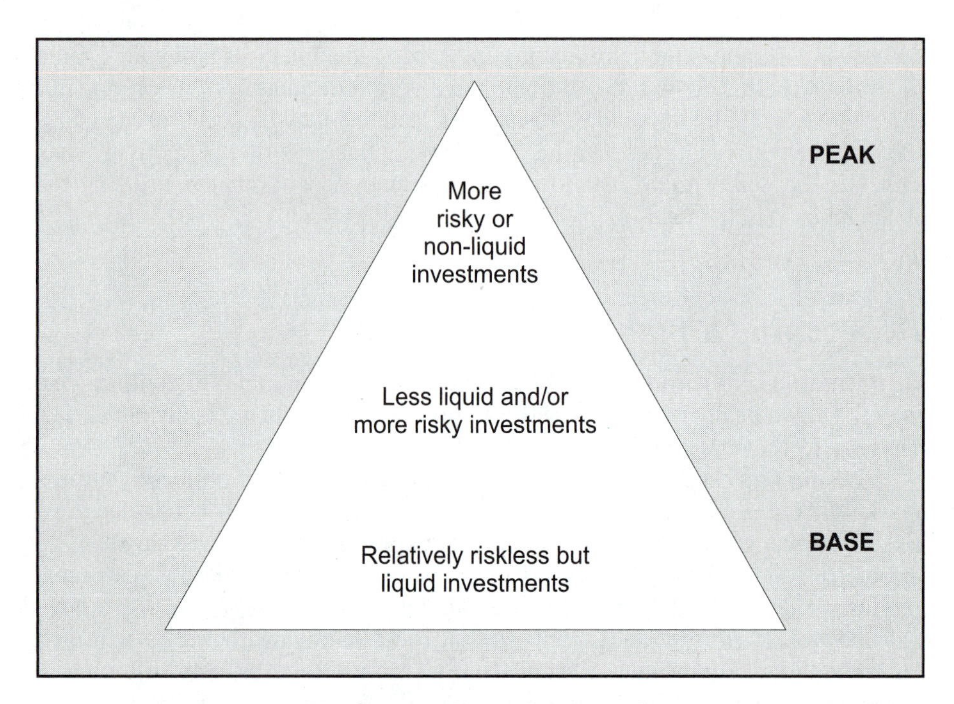

Figure 1.1 Financial pyramid

living expenses are covered and your growth investments are still intact. Thus, the pyramid reinforces the "can afford to lose" concept because each layer is smaller than the preceding base. If all your nest egg and growth funds were invested in the opposite fashion, where the risky portion was on top and the largest as indicated by the inverted pyramid of Figure 1.2, a turn of events for the worse could wipe out the other layers.

Each person's or firm's financial resources and attitudes toward risk affect the size or thickness of each layer, the total size of the pyramid, and the composition of investments in each category. Real estate may be non-liquid to some person or firm but liquid to others, and some properties may be very risky but others are relatively riskless.

Consider again the risk-loving gold speculator. If the $4,000 investment is his rent and egg money, that is, his pyramid base, he opens the possibility of personal financial hardships. However, if it is $4,000 off the top, the situation is entirely different.

Where do futures and options contracts fit in the financial pyramid? Actually, they can fit at all levels. Futures and options can truly be very speculative investments or very low risk investments if properly planned. They can provide excellent potential for risky investment because of the leverage concept or they can provide more controlled, less risky growth through spreading, hedging, or proper use of options. Hedging can protect the risk-averse individual from major

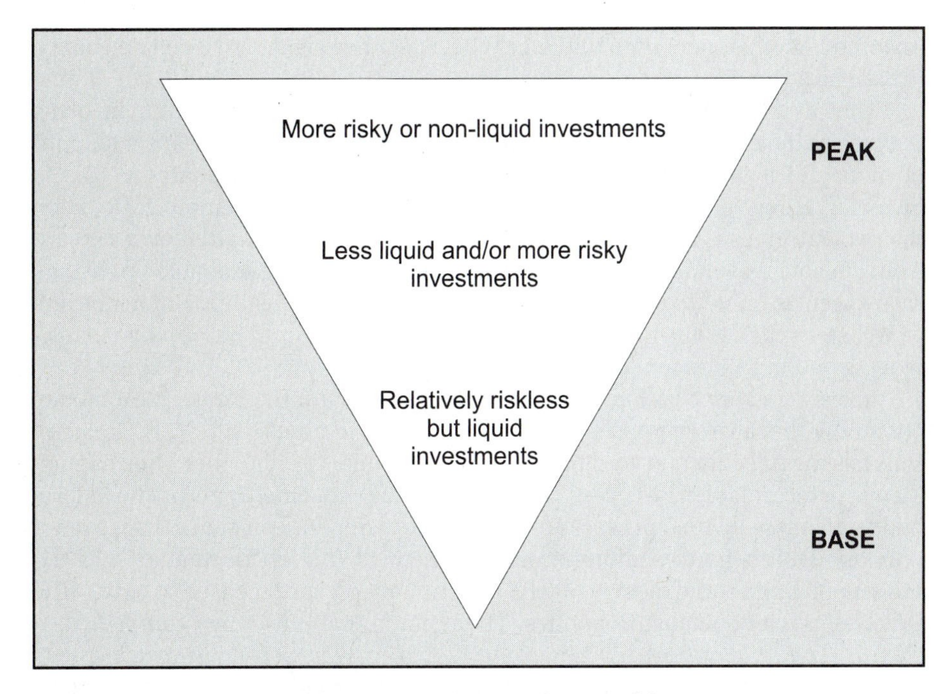

Figure 1.2 Inverted pyramid

price risks. In other words, futures and options can provide many investment opportunities, but they must be evaluated as part of an overall plan with definite goals, attitudes towards risk, and financial resources identified.

A Word about Investing

> ***invest*** *vb* [Fr. L. to clothe] *vt* **1:** to commit (money) in order to earn a financial return **2:** to make use of for future benefits or advantages *(Webster's New Collegiate Dictionary)*

Many people believe that if they buy $1,000 worth of stock in a particular company and over the course of a year the stock declines in value to $500, that their investment was not very good. But they believe (rightly so) that it was still an investment, even though it lost money. They consider buying stock in a company an investment rather than gambling because companies are supposed to make a profit and stay in business, not have losses and go out of business. There is a reasonable chance of earning a return on money committed to buying stock in a company. Yet most reasonable people would not call the $1 used to buy a lottery ticket for a chance at $20 million an investment—even though it meets the strict definition outlined by Webster. They would simply say it was gambling (Webster's says: “to bet on an uncertain outcome”) because there isn’t a reasonable chance at earning a return.

Thus, it seems that the popular view of investing versus gambling involves the idea of the expectation of the financial return from committing a set amount of money. If it is reasonable to expect a return, though not guaranteed, it is investing. However, if the chance for a return is remote, it is gambling. Between the two extremes is a term called speculation—money committed on a venture from which a reasonable return is possible but only slightly probable (Webster’s says speculation is “to assume a business risk in hope of gain”). But according to Webster’s the $1,000 investment in company’s stock could be either an investment, gambling, or speculation!

Investing cannot be limited to ventures that guarantee a return on money committed because there is no such thing. Money committed to U.S. Treasury Bills has the full faith and credit of the U.S. government behind them, and as such the expected default rate is very close to zero—but not exactly zero. The federal budget impasse of 1995 proves that the U.S. government can shut down, even if only briefly. It behooves all investors to remember that on December 31, 1991, the second largest and most powerful government on earth ceased to exist—the Union of Soviet Socialist Republics. The point is that *all money committed to earn a financial return carries some risk of either not earning what was expected or a loss.* A more correct definition of “invest” would be: to commit (money) to earn an *uncertain* financial return; to make use of for *uncertain* future benefits or advantages (**syn:** speculate, gamble).

This definition allows for a more precise explanation of how people use money. One thousand dollars invested with a professional poker player has a higher probability of earning a financial return than does the same $1,000 randomly given to a stranger at the poker table. Both are investments—but with different uncertainties of financial returns. By the same token, $1,000 given randomly to a stranger on the street to buy stocks in U.S. companies is a different investment than the same $1,000 placed with Fidelity (the world’s largest investment firm with a proven track record of stock investments). To label all money placed on the poker table as gambling is foolish, as is the belief that all money placed in the stock market is a safe investment. In truth, the professional poker player and a professional mutual fund manager might yield similar results with similar levels of risks.

The modified definition of investments is used throughout this book to emphasize the importance of viewing all money commitments as investments with uncertain outcomes. Furthermore, the second definition given in Webster’s (with the before mentioned modification) of investing “to make use of for future *uncertain* benefits or advantages” is one of the most powerful uses of futures, options, and other derivatives. Futures, options, and derivatives can help individual investors and businesses have future benefits/advantages by controlling the risk of price changes, increased profits, and established price floors and ceilings for business contract purposes.

PROBLEMS

1. George believes that gold prices will go up, so he buys 100 ounces of gold for a price of $350 per ounce. Two days later, the price of gold has increased to $353 per ounce, so he buys another 200 ounces. Three days later the price of gold has increased to $356, and George buys another 300 ounces. One day later, the price of gold drops to $354 per ounce, and George sells all 600 ounces at that price. What was the net profit (loss) to George? What would have been the profit (loss) situation if George had constructed a normal pyramid (i.e., initially buy 300 ounces, then 200 ounces, and finally 100 ounces)? Explain the difference. Now assume that George gets out at a price of $358 per ounce rather than at $354 per ounce. Which is better from a profit standpoint, the inverted pyramid (100 ounces, then 200 ounces, and finally 300 ounces) or a normal pyramid (300 ounces, then 200 ounces, and finally 100 ounces)?

2. You have $1,000 to invest and only two choices: (1) Place the money in a savings account at a bank and earn 4%, or (2) Place the money as a down payment on a vacant lot next to your current residence (total price $10,000 less $1,000 down with the remainder financed). What factors would you need to consider in making the choice? Explain.

Mechanics of Trading

"The substance of the eminent socialist gentlemen's speech is that making a profit is a sin. It is my belief that the real sin is taking a loss."

—Winston Churchill

Many investors avoid the futures markets because they think that futures are complicated and risky. It is true that futures can be risky, but they do not necessarily have to be risky; and further, they really are not all that complicated. Like many professions and other markets, futures markets have their own language, a set of terms that at times seems to have been created solely to confuse people. But, once some simple rules and terminology are learned, futures markets lose their complicated appearance.

Futures markets are a lot like the more challenging games such as Chess, Backgammon, or Kensington: The rules are very simple and easily learned, but you can spend a lifetime perfecting your strategies. It is this aspect of futures markets that makes them interesting investments. They can provide very risky, highly leveraged investments or very low-risk investments. Futures markets attract investors ranging from high rollers to farmers trying to protect a very thin profit margin. Despite the goals of the investor, the basic concepts are the same and are easily learned. As with all investments, the more time spent studying futures markets, the greater the probability of successful investing. It has been estimated that a Chess Master spends ten years and at least ten thousand hours learning how to master the game of chess. Being a master investor in futures and options markets requires the same dedication, concentration, study, and effort.

The Basics

Simply stated, a **futures contract** is a binding obligation to purchase or sell a specific kind and amount of a particular commodity (or item) at a specific location and time. The price of that contract is determined by **open outcry** auction at a centralized exchange.

The major exchanges are nonprofit corporations and exist to provide the trading place, set rules, and oversee the activities of their members. A member may trade on the floor (or in "the pit" as it is often called) and may buy or sell for himself or act as an agent for someone else and trade for them. Memberships (called "**seats**") are sold only to individuals. Occasionally, new issues of memberships are conducted, such as the new memberships for options trading. But usually, memberships are available only through a secondary market created by current owners who wish to sell their seats to others. The memberships are expensive. A full membership seat on the Chicago Board of Trade or the Chicago Mercantile Exchange costs approximately $1,000,000.

Because seats are expensive and limited in number, the majority of investors trade through brokers, who in turn trade through members of the exchanges. However, since the early 1980s, members have been leasing seats on a monthly basis to individuals. These individuals must go through training exercises just as if they had purchased the seats. Typical leases are for six months payable monthly and can cost at least $5,000 per month for a full seat. Individuals who want to try their skills at trading in the pits find that leasing seats requires fewer financial resources in the short run than purchasing a seat.

Buying, Selling, and Offsetting

Buying in the futures markets is like most investment buying; that is, the investor's goal is to sell later at a price higher than the buying price. But, there is also another kind of action in the futures market. You can sell a futures contract with the anticipation of buying it back at a later date, hopefully at a price lower than the selling price. This action is called selling short (as it is called in the stock market), but most futures traders simply refer to an initial sell in the futures market as a short. Thus, an initial sell is called a **short**, an initial buy is called a **long**.

Therefore, if you buy a futures contract, you are said to be long; and when you sell it later, you have **offset**. You are not short when you offset a long with a sell. You are short only when you initially sell and remain short until you buy it back. Once you have bought and then sold (a buy-sell) or sold and then bought (a sell-buy), you have completed a **roundturn** (an offset).

Don't get hung up on the short idea. Although it may be difficult to believe that you can sell something you don't own, ownership of the underlying commodity is not the issue. The real issue is that you have entered into an obligation or promise very much like a loan to buy a car or a mortgage on a home. Remember, a futures contract is a promise. What you own after you initially buy or sell is simply a contract to do something in the future. When you initially sell a futures contract, you have promised to deliver a specific commodity (item) to a specific place on a certain date. If you buy a futures contract initially, you have promised to accept **delivery**. Because there must always be a trade—that is, for every buyer there must be a seller—the equation always remains in balance. For every buy, there is a sell and vice versa. Although there may be situations in which you wish to perform on your promise by delivering or accepting delivery (although typically you, as with most other traders, will not), you can be relieved of either promise by simply offsetting a buy with a sell or a sell with a buy.

A Word about Selling

Most individuals understand very quickly the idea of buying first and later selling because they all have experience doing it in everyday life. People buy homes and later sell them and likewise do so for cars, stocks and bonds, and other valuables. However, the idea of selling something first and later buying is one of the most difficult concepts (but one of the most powerful) to understand. People say the concept is difficult because they don't have any experience doing it, and thus it is hard to grasp. Surprisingly, many people have had experiences involving the concept but did not recognize its form.

A building contractor who gives a bid on constructing a home for a client is offering to sell a contract to build the home for a set amount with the hopes that she can later buy the materials and construct the home for less than the contract price. The contract that is first sold is for a finished home and they later buy the labor and products to make the finished home, but the process is still an initial sell and a later buy. If the later buying of labor and materials is less than the initial sell, the contractor has made a profit, or vice versa, a loss.

If a neighbor wants your classic 1955 Thunderbird, but you refuse to sell it to her, you might offer her the following contract: I will deliver to you in six months a 1955 Thunderbird that is similar to mine, for $40,000. If she accepts, you have initially sold her a contract that you could fulfill by using your car, but you don't want to. So you shop around to find a similar car to yours during the next six months and buy it, hopefully for less than $40,000. You have completed an initial sell and a later buy.

The process of initially selling a contract and then later buying it is powerful because it allows investors to take advantage of declining markets or inefficient markets that have excess profits. Without this tool the average investor can only profit during market increases, thus knowing how to use the sell-buy tool doubles certain investment opportunities.

Contract Months

Futures contracts call for a specific promise to take place at a specific future point in time. More correctly, they call for a promise to deliver or accept delivery sometime within a specified future month. Futures contracts for a particular commodity generally do not exist for all calendar months, rather, futures contracts will usually be traded for some subset of the twelve calendar months. The futures contracts will usually coincide with some biological or technical condition in the cash market. For example, in corn the contract months are September, December, March, May, and July. September is called "new crop month" because harvest begins to bring in enough corn to establish a liquid cash market. The remaining months exist to provide a storage base for the new month. Theoretically, futures contracts could exist for all calendar months for corn, but the actual trading for some months would be very "thin" and not produce enough volume to maintain a competitive price or liquidity for trading. Consequently, the corn futures market is much better off with five heavily traded, or liquid, months rather than twelve thin, illiquid months.

Similar patterns exist for all the agricultural commodities. The financial contracts (including foreign currencies) are on a quarterly system with futures contracts existing for September, December, March, and June. Likewise, other commodities have months for the futures contracts keyed to volume in the cash market or other technical factors peculiar to that commodity. Certain products such as natural gas, gold, and silver have contracts for all calendar months.

At various times, additional futures contract months have been added to certain commodities on a trial basis and were kept if there was enough volume to justify keeping that month.

The Delivery

If you buy (agree to accept delivery) a futures contract, you can get out of the obligation in either of two ways: (1) accept delivery of the item specified in the contract when the contract expires, or (2) sell an identical futures contract (offset) before expiration. You have bought a promise and then later sold a promise, so you are out of the futures market. For selling and then buying, a similar but opposite process holds.

Because the futures market deals in promises, very little actual delivery takes place. Typically, the futures markets are not cash or spot markets—that is, they do not deal with trading the actual commodity. They don't try to be. You can use them to actually deliver the commodity (item) or a place to acquire the commodity, but most traders avoid the actual delivery through offsetting. Delivery usually never amounts to more than 5 percent of the contracts.

There are times when both speculators and hedgers go through the actual delivery process. These are discussed in later chapters when specific trading strategies are outlined.

You do not need to worry that you will get a truckload of corn or pork bellies dumped on your front lawn. All futures contracts have **delivery points** specified. You will get a delivery notice telling you where the delivery will be in the rare event that delivery actually does take place.

Several contracts avoid the physical delivery process altogether by having **cash settlements**. Both buyers and sellers agree, through the standardized terms of their futures contract, to close their positions (settle), after the final date that futures contracts for a specific month can be traded, in cash rather than by exchanging the physical commodity. All index futures have cash settlements, as well as do feeder cattle and several others. The idea of cash settlement has been around for almost as long as futures contracts, but its first major use occurred with the index futures. Obviously it is difficult to deliver an intangible index, thus cash settlement became necessary. Similarly, feeder cattle were difficult to deliver because of quality variations and multiple production and distribution points.

Prices, Volume, and Open Interest

Unlike the cash market, the futures market limits daily price movements for most contracts. For example, in the cash pork belly market, prices move up or down as much as buying and selling pressure dictates. But, in the futures market for pork bellies, the price can move either up or down from the previous day's settle no more than $1.50 per hundredweight. If yesterday's settle on August pork bellies was $85.00 per hundredweight, then today's trading limits would be $86.50 top and $83.50 bottom. The price of August pork bellies today can trade only within those limits. If it hits $86.50, the market is said to be limit up and no further trading can take place above the limit, likewise for down movements.

Limits help wild speculation from moving the markets too quickly. After a few days of limit trading, the limits are usually widened to accommodate the new price move, and during the delivery month most limits are removed.

Volume is the measure of trading activity. **Open interest** is a measure of open contracts, that is, contracts that have not been offset.

To help explain the concepts of volume and open interest, consider the information in Table 2.1. Mr. A sells to Miss B (Miss B buys). Therefore, the volume is one, and the open interest is one. Mr. A turns around and buys from

Miss X (Miss X sells). Thus, the volume is two, but the open interest is still one. Two contracts have been traded to this point, but only one is open (has not been offset). Mr. A is out of the market because he has offset his position, but the market is left with one contract open: Miss B (buy) and Miss X (sell). In other words, Mr. A entered in the market short, then offset by buying, and is therefore out entirely. The difference in the sell and buy prices at the times when he sold and bought determined whether he made or lost money. In this case, he made $.10 per bushel because he sold at $3.00 and then bought at $2.90.

Table 2.1 Volume and Open Interest Example

Trade	Accumulated Volume	Open Interest	Profit or Loss
Mr. A sells to Miss B (buys) at $3.00.	1	1	
Mr. A buys from Miss X (sells) at $2.90.	2	1	+ $.10 for Mr. A
Miss B buys from Miss G (sells) at $3.05.	3	2	
Miss X buys from Miss B (sells) at $3.05.	4	1	+ $.05 for Miss B – $.15 for Miss X
Miss G buys from Miss B (sells) at $2.95.	5	0	+ $.10 for Miss G – $.10 for Miss B

The example of Table 2.1 shows the zero sum principle of futures markets. For every buy, there is a sell and vice versa. Also, the profit and loss is in balance, $.25 in gains and $.25 in losses. And, the example illustrates one further point, a buyer contracts to buy a specific amount of a specific commodity at a specific point in the future but not from a specific seller. The futures contract does not specify names of the two parties (except for record keeping and accounting purposes).

The volume and open interest figures are usually published along with daily prices. Table 2.2 is a sample of a typical quotation listing from the *Wall Street Journal* for Thursday, July 10, 1997. Take, for example, soybeans. It is the Chicago Board of Trade contract (CBT) for 5,000 bushels and the price is quoted in cents per bushel. The November 1997 contract opened trading at $6.02 and 1/2 per bushel, reached a high during the day of $6.06 and a low of $5.93, and settled towards the end of trading for the day at $5.96 and 1/4. The change from the previous day's settle was a –$.11 and 1/4. In other words, this day's settle was $.11 and 1/4 per bushel lower than the previous day's. Since the November contract has been trading, it has reached a high of $7.50 and a low of $5.77 per bushel. The number of open contracts (those that have not yet been offset) is 68,988 for the previous day's trading (July 9, 1997).

Table 2.2 ***Wall Street Journal*** **Sample Quotation Listing for Thursday, July 10, 1997**

FUTURES PRICES

Thursday, July 10, 1997

Open Interest Reflects Previous Trading Day.

						Lifetime		
	Open	High	Low	Settle	Change	High	Low	Open Interest
				GRAINS AND OILSEEDS				
CORN (CBT) 5,000 bu.; cents per bu.								
July	249	249	247	247 3/4	– 2	393	240	16,087
Sept	236	237 1/2	233 1/2	234 1/2	– 2 3/4	335	227 1/2	62,928
Dec	235 1/2	236 1/2	236 1/2	234 1/2	– 2 3/4	310	227 1/2	150,343
Mr98	243	243 1/4	243 1/4	241 3/4	– 2 3/4	305	236	26,892
May	248 1/4	248 1/2	248 1/2	247 1/4	– 2 1/4	303	241 3/4	4,584
July	251 3/4	252	252	250 1/2	– 2 1/2	315 1/2	245	8,208
Sept	248	248	248	246 1/2	– 2 1/2	260 1/2	244	762
Dec	252	252 1/2	252 1/2	251	– 1 3/4	293	247	4,523
Est vol 40,000; vol Wed 70,083; open int 274,396, –1,671.								
OATS (CBT), 5,000 bu.; cents per bu.								
July	164 1/2	165 1/2	164 1/2	165 1/2	.....	223	148 1/4	416
Sept	148 1/2	148 1/2	147 1/4	147 1/2	– 1/2	185	144	2,511
Dec	146 1/4	146 1/4	145 1/4	146	– 1/4	1834	143	5,109
Mr98	149 1/2	150	149 1/4	149 3/4	– 1/2	174 1/2	148 1/2	563
Est vol 700; vol Wed 1,045; open int 8,616, +124.								
SOYBEANS (CBT) 5,000 bu.; cents per bu.								
July	796 3/4	797	760	779	– 10	902	611	4,488
Aug	749	749 1/2	729	746	– 2	869 1/2	663	33,265
Sept	649	652 1/4	637	645 1/4	– 5 1/4	803 3/4	605	12,687
Nov	602 1/2	606	593	596 1/4	– 11 1/4	750	577	68,988
Ja98	605 1/2	605 1/2	597	600	– 11 3/4	752	583	13,751
Mar	619 1/2	619 1/2	606	608 1/4	– 11 1/4	749	593	3,679
May	620	620	613	617	– 9 1/2	745	601	2,722
July	622 1/2	625	622	623	– 11	751	611 1/2	1,440
Nov	606	615	606	612	– 9	702	597	906
Est vol 48,000; vol Wed 72,083; open int 141,931, +943.								
SOYBEAN MEAL (CBT) 100 tons; $ per ton								
July	263.00	263.00	256.50	260.80	– .90	297.80	199.20	7,049
Aug	246.00	246.00	239.50	243.80	.....	283.80	203.50	25,079
Sept	223.30	223.50	219.30	221.50	– 2.20	262.50	201.50	17,100
Oct	206.00	206.00	202.00	203.60	– 3.60	240.00	193.00	13,821
Dec	199.50	199.50	193.50	194.30	– 4.10	234.00	186.00	36,135
Ja98	195.00	196.00	192.30	192.70	– 4.00	230.00	185.50	4,304
Mar	195.00	195.00	190.50	190.80	– 3.70	227.00	184.50	7,096
May	194.50	198.20	190.50	190.50	– 3.70	222.00	185.50	2,526
July	195.00	195.00	192.00	191.50	– 3.00	217.00	188.50	818
Est vol 21,000; vol Wed 0,257; open int 114,075, –2,296.								
SOYBEAN OIL (CBT) 60,000 lbs.; cents per lb.								
July	22.13	22.13	21.75	21.84	– .22	29.10	21.37	2,480
Aug	22.25	22.25	21.81	21.89	– .30	28.90	21.50	24,866
Sept	22.33	22.33	21.93	21.97	– .34	28.45	21.70	14,403
Oct	22.31	22.35	21.97	21.99	– .34	27.70	21.76	14,353
Dec	22.43	22.43	22.06	22.08	– .36	27.50	21.85	41,091
Ja98	22.50	22.50	22.20	22.21	– .31	27.45	21.98	4,861
Mar	22.78	22.78	22.48	22.51	– .29	27.50	22.20	2,578
May	22.75	22.81	22.61	22.61	– .34	27.55	22.35	1,713
Est vol 17,000; vol Wed 28,893; open int 106,878								

(continued on the following page)

Table 2.2 *Wall Street Journal* Sample Quotation Listing for Thursday, July 10, 1997 *(concluded)*

FUTURES PRICES

Thursday, July 10, 1997

Open Interest Reflects Previous Trading Day.

	Open	High	Low	Settle	Change	Lifetime High	Lifetime Low	Open Interest
Aug	19.42	19.52	19.07	19.22	– 0.24	22.75	16.88	78,786
Sept	19.59	19.66	19.23	19.38	– 0.21	22.30	16.71	58,609
Oct	19.72	19.72	19.35	19.45	– 0.21	21.87	16.84	37,478
Nov	19.73	19.73	19.40	19.52	– 0.20	21.60	16.90	21,009
Dec	19.50	19.62	19.45	19.56	– 0.19	21.46	16.80	43,427
Ja98	19.60	19.61	19.54	19.60	– 0.18	21.33	17.04	21,947
Feb	19.58	19.64	19.50	19.63	– 0.18	21.24	17.15	10,042
Mar	19.55	19.67	19.55	19.65	– 0.18	21.14	17.30	5,515
Apr	19.68	19.68	19.68	19.67	– 0.18	21.10	17.38	4,927
May	19.70	19.70	19.70	19.69	– 0.18	20.95	17.39	7,094
June	19.62	19.72	19.62	19.69	– 0.18	20.90	17.17	25,467
July	.....	.....	.....	19.68	– 0.18	20.65	17.60	6,932
Aug	19.65	19.65	19.65	19.67	– 0.18	20.57	18.80	12,223
Sept	19.63	19.70	19.63	19.66	– 0.17	20.55	17.94	3,426
Oct	.....	.....	.....	19.64	– 0.17	20.55	17.75	1,859
Nov	19.60	19.60	19.60	19.63	– 0.17	20.53	19.22	986
Dec	19.60	19.66	19.55	19.62	– 0.17	20.65	17.05	18,857
Ja99	19.60	19.60	19.60	19.62	– 0.16	20.13	17.85	11,020
Feb	19.60	19.60	19.60	19.62	– 0.16	20.32	19.28	5,467
Mar	19.60	19.60	19.60	19.62	– 0.16	20.14	18.70	1,532
Apr	.....	.....	.....	19.63	– 0.16	20.27	19.60	1,376
May	.....	.....	.....	19.64	– 0.16	20.29	19.82	225
June	19.61	19.61	19.61	19.65	– 0.16	20.47	17.80	5,530
July	19.70	19.70	19.70	19.66	– 0.16	20.14	19.70	785
Aug	.....	.....	.....	19.66	– 0.16	.....	.....	100
Sept	.....	.....	.....	19.66	– 0.16	20.00	20.00	685
Dec	19.75	19.75	19.75	19.67	– 0.16	20.75	17.62	14,384
Dc00	19.75	19.75	19.75	19.70	– 0.16	20.75	19.05	2,412
Dc01	.....	.....	.....	19.74	– 0.16	20.98	19.58	3,055
Dc02	.....	.....	.....	19.80	– 0.16	21.38	19.97	1,375
Dc03	.....	.....	.....	19.91	– 0.19	22.00	19.91	1,712

Est vol 108,993; vol Wed 111,698; open int 409,575, +4,456.

HEATING OIL NO. 2 (NYM) 42,000 gal.; $ per gal.

	Open	High	Low	Settle	Change	Lifetime High	Lifetime Low	Open Interest
Aug	.5270	.5285	.4190	.5211	–.0054	.6130	.4860	37,247
Sept	.5330	.5334	.5250	.5279	–.0051	.6110	.4940	23,414
Oct	.5418	.5418	.5335	.5359	–.0056	.6200	.5050	20,053
Nov	.5505	.5505	.5445	.5449	–.0056	.6180	.5075	14,911
Dec	.5590	.5590	.5515	.5534	–.0056	.6205	.5179	15,181
Ja98	.5640	.5640	.5590	.5584	–.0056	.6205	.5240	13,114
Feb	.5655	.5666	.5605	.5609	–.0046	.6150	.5425	6,667
Mar	.5595	.5595	.5545	.5534	–.0041	.5960	.5250	6,303
Apr	.5440	.5460	.5420	.5414	–.0031	.5820	.5350	3,319
May	.5335	.5350	.5325	.5309	–.0026	.5710	.5250	1,500
June	.5300	.5325	.5300	.5284	–.0021	.5625	.5240	2,804
July	.5325	.5345	.5325	.5304	–.0016	.5650	.5265	1,137
Aug	.5395	.5400	.5385	.5359	–.0016	.5650	.5320	344

Est vol 25,249; vol Wed 27,222; open int 146,476, –1.328.

GASOLINE-NY Unleaded (NYM) 42,000 gal.; $ per gal.

	Open	High	Low	Settle	Change	Lifetime High	Lifetime Low	Open Interest
Aug	.5820	.5845	.5730	.5860	+.0037	.6725	.5465	35,024
Sept	.5750	.5765	.5645	.5740	–.0004	.6340	.5435	15,308
Oct	.5550	.5595	.5540	.5582	–.0027	.6265	.5360	8,490
Nov	.5480	.5520	.5480	.5507	–.0035	.6030	.5350	2,860
Dec	.5455	.5470	.5445	.5480	–.0037	.6045	.5330	6,303

Getting Started

To actually trade commodity futures on the exchange floor, the trader must own or lease a membership at the exchange. But, an individual may trade if he knows someone who owns a membership or he may use a broker. The majority of investing in futures contracts is done through brokers who in turn use members of the exchange to process orders.

Selecting a Broker

There are numerous brokerage firms throughout the country. Choice of a broker depends on which services you want the broker to provide and which services you are willing to pay for. If you collect all your own information, don't want or need any advice, then you may want to select a "discount" brokerage firm or individual. They will process your order for a fee and basically provide no other service. Other brokers and brokerage firms offer a full range of services. They will not only process your order, but they will provide publications, information, and advice. They usually charge a higher fee than the discount brokers.

Brokers are paid a fee for getting you in the market and then out. They do not collect a percentage of the price differences. Typical fees vary for a roundturn (opening and closing your position) from $15 to $75. Thus brokers are paid according to how much (how often and/or how many contracts) you trade, not how much money you make or lose. Unscrupulous brokers, therefore, may try to churn your account. They try to get you to invest and trade often, not necessarily wisely.

When selecting a broker, consider these factors:

1. Fees charged. If you want the cheapest deal and no other services, your selection of a broker is easy. If you want other services, then the fee charged will vary. Shop around and get fee schedules and then look at the following other factors.

2. Length of time in the profession. Oftentimes, the longer a broker is in business, the larger the client base and thus the higher the volume of trades. He or she isn't as likely to try and churn your account. On the other hand, brokers just starting out are eager to try and build a base of clients. However, if they are properly trained, they will not churn your account.

3. Knowledge of the market. If you wish your broker to provide information and/or to advise you, make sure the broker candidate knows not only the futures market but the particular commodities or financials you are trading. Length of time in the business may or may not be helpful here. Just because some brokers have been around awhile does not indicate that they have done anything really good. It may only indicate that they have not done anything really bad. Don't be afraid to ask them questions. Ask for information to read. Try to get a feel for how much they really know about the market. Reputable brokers won't mind; in fact, they will encourage your questions.

4. Reputation of the firm. If your broker is part of a brokerage house, check their references. Check with the **Commodity Futures Trading Commission (CFTC)** or the National Futures Association (NFA) for complaints and or problems both with the firms and individuals. Each year, literally millions of dollars are lost by investors to fly-by-night unscrupulous operations.

In other words, do your homework. If you are going to be a serious investor in the futures and options markets, invest some time and energy in selecting a broker. Consider it a fixed cost. The variable costs with a good broker will be minimal. Otherwise, the situation is reversed.

The Margin

To control a futures contract, you must put up a portion of the contract's face value called a **margin**. The margin required will vary among exchanges, brokers, and commodities but usually ranges from 5 to 15 percent of the contract's face value. Sometimes it is less than 5 percent, sometimes all leverage vanishes and it reaches 100 percent. These events are rare and involve unusual circumstances.

Take an example of a gold trader. She buys a December gold futures contract (100 ounces) for $400 per ounce. The face value is $40,000, and the margin required to control the contract is $4,000. Thus, she places a $4,000 margin with the broker. In addition to the responsibility of submitting the original margin, she must maintain the margin level to retain control of the contract. Therefore, as the price of December gold futures changes, so does her margin. If the price increases over the next few days to $410 per ounce, she has accumulated a **paper gain** of $10 per ounce, or $1,000 for the contract. Her margin account with the broker has increased to $5,000 (Table 2.3). She can withdraw her paper profits of $1,000 and leave the margin account at the specified $4,000. This allows her to withdraw most of her paper profits as they accumulate rather than all at once when she finally offsets her long December gold futures contract with a sell.

On the other hand, prices may move against her. If December gold moves down to $395 per ounce, she has a **paper loss** of $5 per ounce or $500 in total. Her margin account is really at $3,500. In addition to the margin level, the broker will specify a **maintenance margin** level. This varies but will usually be approximately 75 percent of the original margin. In the example of the $4,000 margin for gold, the maintenance margin level would be set at say $3,000. This is the level below which the margin account will not be allowed to fall. If the account does fall below this level, additional funds must be placed in the account quickly. The $5 price decline (to $395) or $500 total loss that left the margin account at $3,500 was not large enough to trigger the maintenance margin level of $3,000. She would not be required to place additional funds in the margin account to cover the paper loss. However, if the market continued down say to $388 for another $7 per ounce loss (a contract total loss of $1,200), the margin account would fall to $2,800. This is $200 below the maintenance margin level and, thus, it is now necessary to bring the margin account back to its full face value of $4,000. A **margin call** will be given by the broker to the trader for $1,200.

Table 2.3 Margin Call Example

	Contract Value (100 oz. × Price)	Margin Account (Maintenance margin [MM] set at 75% of initial)
Buys December Gold at $400/oz.	$40,000	$4,000 initial (MM = $3,000)
Price ↑ to $410	$41,000 +$ 1,000	$5,000
	Trader can give broker a margin call to remove the $1,000 paper profit	
		– $1,000 $4,000 Paper Profit
Price ↓ to $395 (Assume that the trader did not give the broker a margin call as above)	$39,500 –$ 500	$4,000 $3,500
	No margin call because MM not reached.	
Price ↓ to $388	–$ 700 $38,800	$2,800
	Margin call of $1,200 to bring margin back to full value.	

The concept of a maintenance margin level is to allow for small price moves that invariably occur, but to keep the margin from deteriorating significantly. The trader is allowed small price moves against her position but is not allowed to let paper losses accumulate excessively.

In addition to the margin required on each contract traded, a broker requires you to show financial solvency. Opening an account with a broker requires some degree of financial liquidity. That is, the broker wants you to have a liquid financial base that can be converted to cash to pay any losses should they occur. Thus, a broker may require you to put up "good faith" money to open an account. This may amount to $5,000, $10,000, or even more depending upon whether you are speculating or hedging and how much you intend to trade. The broker wants to be sure you can meet margin calls. Some brokers require you to actually place cash in a security account with the firm to open the trading account. Others may require assignment of other liquid assets such as Treasury bills or certificates of deposit. Still others may require only that you have the assets in some liquid form. Again, it pays to shop and compare brokers and brokerage firms and their policy on opening accounts, when they will release good-faith deposits, and whether interest is paid on the committed funds.

Types of Orders

Once you have opened an account, you will have to tell the broker what kind of order you want transacted. You must, of course, specify the commodity to be traded, whether the trade is a buy or sell, and the contract month, but there are many variations that must accompany the buy or sell.

One of the major factors of all orders is time. You can issue a simple "kill or fill" order. If any part (such as the amount or number of contracts) cannot be filled, the entire order is canceled. For example, you issued an order, "Buy two December Gold at $450." If only one can be bought at $450, the whole order is killed.

You could place an "open" or "good till canceled" order, which remains in effect until canceled by you or until the contract expires. Other orders include "Good through (date)" or "Time of Day." You specify the latest transaction time, such as "Sell two July Gold by March 20" or "Sell two July Gold at the market by 10:30 a.m." You can also specify orders "on the close" or "on the open." The main thing to remember is that you can specify the time and the date of the orders you place. Don't worry about the terminology of the orders because the broker will translate your wishes to the jargon of the marketplace. For example, you may say to the broker, "I want to buy two July gold contracts at a price of $450 per ounce or better if I can get them today." The broker will issue a day order, "Buy two July Gold at $450."

Another component of all orders is the price. You may issue a simple "market" order, which means to fill the order at the best possible current market price.

You can also specify price with a "limit" or a "market-if-touched" order. A limit order must be specified below the current market price if the order is a buy or above the current market price if the order is a sell. A limit order allows the trader the flexibility to develop a more controllable trading plan. If you issued a "Buy two December Gold at $445 limit" and the market didn't reach $445 or lower, the order will never be filled. The same is true for a "Sell two December Gold at $455 limit" if the price never reached $455 or higher.

A "market-if-touched" is the same as a "limit" order except that they become "market" orders the moment that the price reaches the specified price. Thus, a "Buy two December Gold at $445 limit" is not the same as a "Buy two December Gold at $445 market-if-touched" because the limit order will not be filled except at $445.00 or lower, whereas the market-if-touched will be filled if the market "touches" the $445.00 level. You may get the market-if-touched order filled at $445.25 or higher if the market very briefly touched $445.00 and moved back up slightly because it became a market order.

Another useful order is the **stop**. You can use the stop order to protect profit levels or to limit your losses, called a "stop loss" order. A stop buy order must be above the prevailing market price and a sell below. Let's assume that you have a December gold purchased at $450. The current price of December gold is $458. If you want to protect your $8 profit, you could issue a "Sell one December Gold at $457 stop." The order will be filled as close to $457 as possible once the price has touched the $457 level. On the other hand, you could use the stop loss order to prevent large losses from accumulating. Thus, you could issue a "Sell one

December Gold at the market stop at $459." If the sell was filled at $458 and the market then moved against you to $459, the stop would be triggered and your position would be offset with a buy as close to $459 as possible. Your losses were limited. Problems can arise when the stop loss is set too close to the market price, because small variations around the market price occur and may trigger your stop loss order causing you to be **whip sawed** by the market and your own stop loss orders. A broker can help you gauge the level of price sensitivity for stop loss orders.

The various types of orders can be combined as you become more sophisticated with your investing. Again, don't get hung up on the terminology of orders. Give the information concerning commodity, amount or size of the order, time, price, and exchange to your broker, and let him or her put in the proper terminology for the execution of the trade.

PROBLEMS

1. Mary decides that gold prices are going to decline, so she places a sell order with her broker. The broker requires an initial margin of $5,000 with a maintenance margin of $4,000. The broker got the contract (December) sold at a price of $376 per ounce (contract size is 100 ounces). At what price will Mary get a margin call from her broker?

2. Calculate the volume, open interest, and profit (loss) for the following trades:

	Volume	Open Interest	Profit (Loss)/Unit
A sells to B at $400/unit			
B sells to C at $402/unit			
A buys from H at $403/unit			
R sells to Z at $398/unit			
C sells to R at $399/unit			

   What additional trades would you suggest to get the open interest to zero?

3. Sam has his broker buy a December gold contract for $405 per ounce (100 ounce contract). He posts an initial margin of $5,000 with a maintenance margin of $4,000. He also tells the broker to put a stop loss order in at a price of $402 per ounce. During the next few weeks, gold prices increase to $430 per ounce. Is there any way that Sam could maintain his long position and still protect the paper profits he has accumulated?

A Brief History of Futures Markets

"I'm astounded by people who want to 'know' the universe when it's hard enough to find your way around Chinatown."

—*Woody Allen*

Early Marketplaces

Organized markets have existed since the early Greek and Roman cultures. The Forum in Rome and the Agora in Athens were established as marketplaces, and several additional trading centers were established during the height of both cultures. During the eleventh and twelfth centuries, organized fairs and marketplaces developed throughout Europe. All these early markets relied on cash (or spot) transactions and some form of time sales (future delivery). Most of the transactions were spot, that is, the product was visibly inspected by the buyer, the buyer and seller negotiated a price, and the buyer took immediate possession. However, several sales involved time contracts. The seller did not have the commodity but promised to deliver it sometime in the future. Price was usually agreed upon in advance of delivery.

Of course, not all the time sales transactions at the marketplaces were handled to the satisfaction of all parties. Disputes over quality, quantity, and credit worthiness arose. Acceptable criteria and rules became codified bylaws in Europe in the late 1300s.

Generally the local fairs and marketplaces became more centralized, specialized, and consolidated. By the fifteenth century, two futures-type contracts were in common use:

1. To-Arrive—delivery in a few days with title passing from seller to buyer
2. Forward—delivery in a few days but title is not passed until delivery

The Royal Exchange was established in London in 1570 trading these futures-type contracts, however it fragmented into several smaller exchanges over the years. Likewise, in the early 1700s in Japan the Osaka Rice Exchange became a centralized exchange trading similar contracts with a rigid set of codes of conduct. Both the Royal Exchange and the Osaka Rice Exchange are considered the forerunners to the current futures exchanges.

Early U.S. Futures Marketplaces

Economic Setting

Producers of agricultural products found that when they brought their products to market, the marketplace could not handle all the production adequately. Thus merchants found that they could offer the lowest possible price and receive all they wanted. Later in the year, however, shortages existed, and prices rose accordingly. This was especially true in the great breadbasket of America called the Midwest. Chicago was the principal marketplace, but it lacked adequate storage and transportation facilities in the early 1800s (Chicago was founded in 1833). Because of the wide variance in prices within the year, several things developed more or less simultaneously.

First of all, to-arrive and forward contracts started being used more and more as ways to smooth the flow of the product to market. As delivery was postponed until later time periods, storage facilities had to be erected to compensate for the fact that nature has the crop ready during only one period of time (harvest). This developed a class of country merchants who would buy the crop from the farmers at harvest, store them, and move the products to meet the various dates on the contracts that they had sold to other processors or merchants in Chicago. Moving the products into and out of Chicago throughout the year necessitated better transportation facilities. Thus, all-weather roads were constructed, as well as the Illinois-Michigan Canal that connected the Illinois River and Lake Michigan in 1848.

As usual, not all of this was accomplished smoothly and without problems. Age-old problems of default on the terms of the contracts as to quality, quantity, and credit worthiness of the parties surfaced time and time again. This prompted eighty-two businessmen in 1848 to found the Chicago Board of Trade. This allowed for a place to trade, a set of rules by which to trade, and a system to assure that traders would live up to the terms of the contracts. The Chicago Board of Trade was instrumental in establishing the first comprehensive set of grades and standards for grain trading.

Several markets developed near New York during the mid-eighteenth century. None of them exists today, but they did provide the foundation for the New York commodity exchanges that operate today.

The to-arrive and forward contracts evolved into a form similar to today's futures contracts by the middle of the 1800s. By the late 1800s, the basic concepts of today's futures contracts were well established:

1. Price determined by centralized open outcry auction
2. "Strong" contracts (i.e., can be retraded)
3. Leveraged contracts
4. Use as price risk management and leveraged speculation tools rather than as an actual delivery process for the cash commodity

Options Markets

Options have enjoyed a long history. In fact, their history is older than futures. Early Phoenician merchants used "rights to buy" on grains. Likewise in Holland the tulip trade used "rights to buy" during the early 1600s. Sir Charles Leonard Woolley uncovered the Tell al Muqayyar in 1923 and found countless clay tablets describing transactions in the city of Ur. Among the clay tablets were records of "payments in kind" for taxes with commodities and "rights to buy" certain commodities. Thus, options are at least as old as recorded history (approximately 6,000 years).

Options have been used in the United States for centuries and were often called "privileges." In drawing up the 1932 Securities Act, the attitude was that because it was difficult to know which were good and bad options as a matter of convenience, all options would be banned. However, the Securities and Exchange

Commission subsequently did allow for stock option trading. In the early 1930s corners were attempted on "privileges" in wheat, which focused enough political pressure to cause the total ban on commodity options in the 1936 Commodity Exchange Act.

A catch in the 1936 Act opened up a thriving options market in 1971 in the so-called "international" commodities (sugar, cocoa, coffee, plywood, silver, silver coins, copper, and platinum). The market collapsed amid scandals and high losses in 1973. Amendments to the Commodity Exchange Act in 1974 gave power to the Commodity Futures Trading Commission (CFTC) to regulate all options trading on commodities and futures. By 1978 the CFTC had a total ban on all options.

In 1982, the CFTC started a pilot program on nonagricultural commodities to trade options on futures. Each exchange was allowed one options market on one nonagricultural futures contract. The pilot program was successful and expanded. By 1984 a thriving options market was once again available in the United States. The revived options market has proved to be successful because:

1. Centralized open outcry auction determines premiums.
2. Contracts are "strong" (i.e., can be retraded).
3. Options are on futures, thus tightly regulated and controlled.

Futures and Options Exchanges

All the major futures and options exchanges are nonprofit organizations that exist to provide a place for trading to take place as well as provide rules and procedures for trading. They supervise and enforce their rules and codes of conduct as well as discipline and/or remove those members who violate the rules. In addition, exchanges do research on new contracts and ways to improve existing contracts as well as provide educational programs. Exchanges know that it is important for the public to understand that they meet the needs of the various industries they represent and that they provide a safe and regulated environment for contracts to freely trade. Thus during the 1980s and up to the present, most of the major exchanges have increased their research, education, and rule enforcement efforts.

Most exchanges have a limited number of memberships that are held by individuals. These memberships are allowed to be resold to qualified individuals. Some exchanges allow multiple memberships, that is, more than one individual can own a single membership. During the last few years, most exchanges have allowed members to lease their seat to another individual for certain periods of time.

Chicago Board of Trade (CBT)

Although the records of the Chicago Board of Trade were destroyed in the Great Chicago Fire of 1871, modern day type futures contracts are believed to have been traded on the CBT as early as 1859 and certainly by the end of the Civil War in 1865. The CBT also started the concept of a margin, or "good faith" deposit for contracts. To buy or sell a contract required that a portion of the value (margin) be put up in cash as a measure of good faith to perform as per the contract specifications. Prior to this concept of margin, some individuals would default on the contract if it were to their advantage. The only recourse the other party to the contract had was to sue in court. With the margin concept, the defaulter would lose the margin and could still be sued. Thus, the use of margin was a major step forward for futures exchanges.

The CBT was the first exchange to establish a **clearinghouse** for futures trades in 1926. The clearinghouse guarantees the financial integrity of the futures contract. Other exchanges quickly followed by establishing separate clearinghouses from the exchange itself. No futures contracts have been in default to the point that any trader has lost money due to a default by the other contracting party since the clearinghouse concept was established.

During the early years of the CBT, all business was focused on the grains (oats, wheat, and corn). Later, after World War II, several other contracts were added such as iced broilers, plywood, gold, silver, and in October 1975 the first mortgage interest rate contract, the Government National Mortgage Association (GNMA). Current futures contracts exist for corn, oats, wheat, rough rice, anhydrous ammonia, diammonium phosphate, gold, silver, U.S. Treasury bonds, U.S. Treasury notes, long term municipal bond index, 30 day fed funds, and PCS Catastrophe Insurance Options (Internet address: www. cbt.com).

Chicago Mercantile Exchange (CME)

The Chicago Mercantile Exchange began in 1898 as the Chicago Butter and Egg Board. The Chicago Butter and Egg Board was officially changed to the Chicago Mercantile Exchange in 1919 and started that same year to trade the first futures contracts on butter and eggs. The CME later tested contracts on onions (outlawed in 1958), Idaho potatoes, scrap iron, turkeys, and apples to name a few. They successfully added live cattle in 1964 and hogs in 1966.

The CME started the process of world futures trading in 1984 when they formed a link with the Singapore International Monetary Exchange to trade currencies and the Eurodollar. They stepped up the global pace again in 1992 by putting the system GLOBEX on line for worldwide electronic trading.

The CME formed the International Monetary Market (IMM) in 1972 to trade foreign currencies. Later gold was added as well as several interest rate contracts, stock index contracts, and options on futures. Both the Merc and the IMM currently trade nearly fifty different futures contracts on boneless beef, fluid milk, cheese, butter, lean hogs, live cattle, pork bellies, feeder cattle, oriented strand board, random length lumber, Euro time deposit, treasury bills, one month Libor, one month fed funds, several Brady bonds, S&P stock index, Nikkei stock index, Goldman Sachs commodity index, major market index, and most major currencies just to name a few of the many contracts (internet address: www.cme.com).

Midamerica Commodity Exchange (MCE)

The Midam got started soon after the Civil War ended in 1868. It specialized in "mini" contracts—contracts similar to other exchanges' contracts but smaller. For example, they trade the same grain contracts as the CBT but for 1,000 bushels rather than 5,000. Likewise for live cattle (20,000 lbs. for the MCE versus 40,000 lbs. on the CME). In 1984 they formed an association with the New Orleans Rice and Cotton Exchange (NORCE). The NORCE was originally founded in New Orleans in 1981 but moved to Chicago in late 1983 and formed the present association with the Midam. It is now known as the Chicago Rice and Cotton Exchange (CRCE). In 1986 the Midam and the CRCE merged with the Chicago Board of Trade. The merger is attempting to maintain the individual status of the Midam and the CRCE yet provide some of the benefits of the CBT's organization.

The Midam trades contracts in corn; oats; wheat; soybeans; soybean oil and meal; live cattle; gold; silver; platinum; Australian dollar; Canadian dollar; British pound; Deutsche marks; Japanese yen; Swiss franc; Eurodollar; U.S. Treasury bonds, notes, and bills (Internet address: www.midam.com).

Minneapolis Grain Exchange (MPLS)

The MPLS is unique among futures exchanges in that it trades spring wheat futures and options, and is also a large cash wheat market. The MPLS was organized in 1881. In 1987 the MPLS added a contract on high fructose corn syrup (later dropped). The exchange trades barley, white wheat, spring wheat, white shrimp, and black tiger shrimp (Internet address: www.mgex.com).

Kansas City Board of Trade (KCBT)

The KCBT got its start in 1856 but it wasn't incorporated until 1973. It trades hard red winter wheat (the major wheat of the Great Plains) and innovated the early grain sorghum futures (no longer traded). The KCBT started the first stock index futures, the Value Line in 1982 and added a mini value line index later. The KCBT started trading western natural gas futures in 1995 (Internet address: www.kcbt.com).

New York Mercantile Exchange (NYME)

The NYME was founded in 1872 to trade eggs, butter, and cheese. The NYME has been the major force in developing energy-based futures and options. It now trades platinum, palladium, silver, gold, copper, crude oil (light, sweet, and sour), two unleaded gasoline contracts, heating oil, propane, electricity, crack spreads, three natural gas contracts, and EuroTop 100 index (Internet address: www.nymex.com).

New York Cotton, Citrus, Finex and NYFE Exchange (NYCE)

The NYCE was formed by a group of cotton merchants in 1870. The citrus division was added in 1966 to trade frozen orange juice concentrate futures and options. Financial futures and options contracts were added in 1985. With the affiliation in 1988 of the NYFE a very broad array of financial futures and options are available. The NYCE currently trades cotton, frozen orange juice concentrate, potatoes, New York Composite stock index, Emerging Market Debt index, 2YTN5YTR Treasury notes, CRB Bridge Futures Price index, eight cross rate contracts, and twelve paired currency contracts (Internet address: www.nyce.com).

The New York Coffee, Sugar and Cocoa Exchange (CSCE)

The Coffee Exchange started in 1882 with sugar added in 1914, and in 1979 they merged with the New York Cocoa Exchange (founded in 1925). Current contracts traded include Coffee "C", Brazil Differential, world sugar no.11, domestic sugar no. 14, world white sugar, cocoa, milk, cheese, nonfat dry milk, butter, and BFP milk (Internet address: www.csce.com).

Other Exchanges

Futures and options are traded on several exchanges around the world. Great Britain, France, Japan, Australia, Singapore, and Canada all have thriving and viable futures exchanges. Additionally, China, Mexico, Italy, Spain, Sweden, Germany, and Belgium (to name only a few) have futures and options exchanges. As trading in cash goods and services worldwide continues to be more global, the need for additional exchanges will no doubt surface in other countries. In fact, one of the reasons that the Chicago Mercantile Exchange developed GLOBEX was to aid in having 24-hour around-the-world trading opportunities for speculators and hedgers. The world of futures and options trading continues to expand rapidly.

Regulating Groups

Commodity Futures Trading Commission (CFTC)

The CFTC was formed in 1974 to replace the Commodity Exchange Authority (CEA). The change in name reflected the changing role of futures and options markets from being dominated by the traditional agricultural contracts to financials, currencies, metals, and energy contracts.

The CFTC has the broad authority to regulate all futures and options trading including but not limited to the contract markets (exchanges), traders, and the terms of trading—thus, all activities involved with futures and options contracts. The CFTC does not regulate cash and forward contract markets except as they relate to manipulation of the futures markets.

The CFTC and the Securities and Exchange Commission's (SEC) duties were outlined in the 1982 Futures Trading Act. The CFTC's primary responsibility is regulation of all futures and options contracts including stock index futures and options contracts. However, the 1992 reauthorization of the CFTC added the Federal Reserve Board (FED) as a new regulator for stock index margins. The FED must approve all margin and exchange rules as they relate to stock index futures and options. The CFTC and SEC have an uneasy partnership and a very difficult job in lines of authority. It is becoming more difficult to separate futures, options, and securities investments especially as countless new derivatives are developed. No doubt in the future, the CFTC, SEC, and FED will share more and more of the responsibility for regulation in the increasingly complex futures and options markets (Internet address: www.cftc.gov).

National Futures Association (NFA)

The NFA is a self-regulatory association formed in 1982. The CFTC requires most people associated with futures and options to be registered with the CFTC and be a member of the NFA. The NFA provides standards of business ethics and screening for potential members. They also provide education and testing for both members and nonmembers (Internet address: www.nfa.futures.org).

PROBLEMS

1. How did the early forerunners of futures contracts (to-arrive and forward) help to develop better transportation and storage facilities?
2. What are the major benefits of having futures exchanges?
3. What role does the Commodity Futures Trading Commission play in regulating futures and options contracts?

Investing as a Speculator—The Basics

"Business is never so healthy as when, like a chicken, it must do a certain amount of scratching for what it gets."

—Henry Ford

Speculators in futures markets do not own or control, nor do they ever really intend to own or control, the underlying commodity or financial instruments. They invest in futures markets to try and capture profit from price movements and/or price relationships. The major attraction of speculative investors to the futures market is the leverage made possible by the margin system and the liquidity of the market that allows the investor to offset his position and thus remove any further risk.

There are three major ways in which to invest as a speculator in futures markets: short term, long term, and spreading.

Short Term

Investors who trade on a daily or within-the-day basis attempt to profit from the fluctuations that arise between the open and the close on any given day. The most celebrated of all day traders is the **scalper** (scalpers are also known as locals as a way to soften the negative image of the term scalper). Locals are almost always exchange members. They trade on very small price movements and concentrate on a large volume of trades to generate investment income. Locals usually end the day of trading without holding any open position. That is, they offset all trades by the end of the day.

Locals provide the necessary liquidity to keep futures markets functional. Although locals are often berated as nonessential to the market, their investing provides the opportunity for other speculators and hedgers to invest. Without the locals' willingness to accept the very short-term price risks, the market would almost certainly lack the necessary level of liquidity to function smoothly. Thus other speculators or hedgers would have a more difficult time getting their trades transacted on the desired schedule.

An example of a local trading short term is when the local buys a December gold at $450.25 per ounce with the hope of selling it quickly and, of course, above his buying price. If he sells it two minutes later at $450.50, he has gained $.25 per ounce, or $2.50 per contract. Though this is not much of a gain on one contract, the local will usually trade several contracts, and you must remember that the local was holding the investment for only two minutes. Speed and volume of trades are the major components of locals' investment plan.

Other than locals, there are investors who trade on day-price relationships such as the difference between the open and close, or those who attempt to ascertain the lows and highs for the day. Many traders may or may not be exchange members. Traders outside of the pit with access to price reporting services day trade.

A **day trader** other than a local may trade the open by selling December Treasury bills and buying them back sometime during the day when the price has moved down to the level the trader thinks is the lowest price for the day. Day traders who are not exchange members must make sure the price move covers their brokerage costs. Exchange members have the benefit of making their own trades and paying only the exchange fee (usually less than $2 per roundturn).

Day traders concentrate on quickness and volume. Furthermore, they seldom let egos get in their way and are usually quick to admit they often make mistakes on price direction and switch to take an opposite position. Day trading thus becomes a fast investment with quick rewards and even quicker losses. Veteran pit traders delight in new exchange members they dub "fresh meat." New members quickly learn the setting or leave, as many end up doing.

Long Term

The length of time that a futures contract is held can be a few seconds or several months. Despite an obvious wonder about the use of the term "long term," for classification sake, any trade held longer than a day is considered a long-term investment. These traders are also known as position traders.

Investors who try to profit from price movements that take longer than a day to form will invest with the anticipation of offsetting when the price movement ends. Unlike day traders who concentrate on small price movements with high volume of trades, long-term traders will concentrate on a smaller trading volume with larger price moves and are also known as position traders.

Position investment in futures markets does not necessarily require access to a daily price reporting service. Rather, long-term investment involves serious study of price movements over time.

Spreaders

Short- and long-term investing is concerned with price movements. Spreading involves price relationships, often over time within the same commodity or instrument, or between closely related instruments. If the price of Treasury bills is increasing, what is the price of Treasury bonds doing? When gold prices decline, do silver prices follow? How soon? How closely? How much? Spreaders try to estimate these relationships and then invest accordingly. Spreading, therefore, consists of holding positions in at least two markets at the same time.

As an example, consider a storable commodity such as corn. The difference between a March corn futures price and a May corn futures price should be the carrying charge for corn between March and May. If the average carrying charge is $.03 per bushel per month, then the normal relationship between the March and May futures prices should be $.06. Investors would watch this relationship; if it increased to $.08 they would "put on a **spread**" by buying the March contract and selling the May contract. When the market corrected itself and went back to the normal relationship ($.06), they would "lift the spread," as illustrated in Table 4.1. The buying pressure on March futures caused the price to increase, and similarly the selling pressure on May futures caused the price to decline, thus driving the spread back to normal.

Table 4.1 Spreading Relationship on Corn Futures

Date	March Corn Futures Price	May Corn Futures Price	Spread
February 1	$2.80	$2.86	.06
February 2	$2.79	$2.87	.08

When the spread widened .02 beyond the normal relationship, a spread investment is possible by Buying March futures at $2.79 and simultaneously Selling May futures at $2.87.

Date	March Corn Futures Price	May Corn Futures Price	Spread
February 3	$2.80	$2.86	.06

Once the spread relationship returns to normal, the spread investment should be lifted by Selling March futures at $2.80 and Buying May futures at $2.86.

Buy March at $2.79, Sell March at $2.80 = .01 Gain
Sell May at $2.87, Buy May at $2.86 = .01 Gain
.02 Net Gain

A similar situation exists when the spread is narrower than normal. An investment to capture this relationship would be to put on a reverse spread and lift it when the relationship returned to normal, as shown in Table 4.2.

Table 4.2 Reverse Spreading Relationship on Corn Futures

Date	March Corn Futures Price	May Corn Futures Price	Spread
February 1	$2.80	$2.86	.06
February 2	$2.81	$2.85	.04

When the spread narrowed .02 below the normal relationship, a reverse spread investment is possible by Selling March futures at $2.81 and simultaneously Buying May futures at $2.85.

Date	March Corn Futures Price	May Corn Futures Price	Spread
February 3	$2.80	$2.86	.06

The spread relationship has now gone back to normal, and the reverse spread investment should be lifted by Buying March futures at $2.80 and Selling May futures at $2.86.

Sell March at $2.81, Buy March at $2.80 = .01 Gain
Buy May at $2.85, Sell May at 2.86 = .01 Gain
.02 Net Gain

The selling pressure on March futures depresses the price, and the buying pressure on May futures increases the price, thereby driving the spread back to normal.

Spread investing of this type is relatively riskless because gains made on either the buy or sell side are usually offset by losses on the other side. The risk is that the two won't match exactly, but that in itself is a much smaller risk than holding either one of the two contracts. The margin requirements for spreads are much lower than for short- or long-term speculation because of the lower risk element.

Spreaders will spread temporal, spatial, form, and substitutional relationships. Temporal relationships involve carrying charges such as storable commodities (corn, soybeans, wheat) or the **cost of carry** for financial instruments. The price differences between trading places such as New York futures gold and Chicago futures gold are spatial relationships that also offer spreading opportunities. The soybean futures market offers form price relationships because the raw product (soybeans) has a futures contract and the major products derived from soybeans—meal and oil—both have futures contracts. Similarly, futures contracts exist for hogs and one of the primary products from hogs—bacon (otherwise known as pork bellies)—as do futures contracts for crude oil and some of its derivative products such as gasoline, propane, and heating oil.

Substitutional price relationships are those commodities and financials that, in a large sense, "substitute" for each other. Certainly, Treasury bills and Certificates of Deposit have some degree of substitutability between each other. Substitutional price relationships are difficult to estimate and pose problems in estimating normal relationships.

If all this sounds too good to be true, the catch is in determining what is normal. Markets are constantly changing, and consequently it is difficult to determine what normal relationship values should be. Sound investments in spreading futures contracts require vigilant, ongoing research effort concerning theoretical and empirical market relationships and prices.

Problems

1. What is the major difference between a position trader and a spreader?

2. Speculators in the futures markets enjoy a major benefit over most speculators in the cash (spot) market. What is it?

3. Judy has observed that the normal relationship between the November gold futures price and the December gold futures price is about $3 per ounce. Today the price of November gold opens at $456 per ounce and the December at $457 per ounce. What should Judy do? Explain. What has to happen for Judy to make money? What are the risks that Judy faces?

Speculative Strategies

"Efforts and courage are not enough without purpose and direction."

—John F. Kennedy

Speculators provide vitality in futures markets. Most traders are speculative traders—90 percent or more of all trades are actually speculative, as opposed to hedging, in nature. Thus, speculators provide liquidity for the marketplace. Without speculators, the markets would be "thin"—not enough buyers and sellers would participate to establish a true competitive price, nor enough volume of trades to allow easy entry and exit.

Needless to say, speculators are very important to the futures and options markets and constitute the majority of volume. Read what Victor Niederhoffer wrote in *The Education of a Speculator* (pages vii–viii):

> I am a speculator, and my daily bread depends on reversing big moves. In economic terms, my function is to balance supply and demand. I sell when prices are high and buy when prices are low. When prices are too high and consumers want to exchange cash for goods, I take their cash and let them have their goods. I prevent shortages by pushing prices down so consumers don't have to pay up. Conversely, when prices are down and producers want cash badly for their goods, I give them the cash and take their goods. In these bad times I keep producers from going broke, and prevent waste and spoilage by bringing prices up. I'm like a dynamic refrigerator, or a captain rationing food on an unexpectedly long voyage.

Yet what speculators do and how they make money is not as easy to describe as the role their actions have on the marketplace. There are some fundamental concepts, but they do not produce results consistently for everyone. If there were truly a strategy that always worked, then everyone would eventually be using it and either it would continue to produce profits for everyone, and thus defy all known laws in economics, or all profits would finally be bid away and the strategy would no longer be profitable.

The successful speculator Victor Niederhoffer says it best in *The Education of a Speculator* (page ix):

> I don't intend to unload any of my secret money-making systems here, for readers' good as well as mine. If I did hold an "open sesame" to the markets, I wouldn't share it. There is ample opportunity to use wealth in this world, and neither I nor my friends, nor anyone else I have ever met, has so much of it that they are interested in putting themselves at a disadvantage by sharing their secrets.

When anyone offers their secrets of making money as a speculator, it would be wise to consider that they are probably offering what once worked but is no longer effective and/or now it is offered to the public and is no longer a secret. In either event, the value of the secret is zero and could be negative.

Although there are no strategies that consistently produce profitable trades, there are some generally accepted speculating behaviors. Each exchange, commodity market, and trading group has its own buzzwords, but the concepts are universal. The three major behavioral concepts are fear and greed, puking, and discipline.

Fear and Greed

Fear and greed are said to be the major reasons why orders are placed or lifted. Consider a trader who has placed a buy order on pork belly futures and the prices start up. Greed takes over and the trader adds to his position—and continues to add as the market moves up (forming an inverted pyramid). Then the market takes a sharp downturn, all profits are erased, and a small loss has accumulated. Now fear takes over, and the trader liquidates all trades at a considerable loss. Greed is the characteristic that causes speculators to add to profitable positions—and indeed often add to losing positions with the thought that they must "make back their losses." Anyone who has spent anytime at the gaming tables in Las Vegas, Reno, or Atlantic City has witnessed the gambler who has abandoned his original game plan (if he ever had one) and "bets it all" hoping and praying that he will win back the car, the house, or just the plane ticket home.

Greed is one of the strongest aspects of human behavior, and therefore it quickly is translated into individual trading styles. It is, of course, the characteristic that causes leveraged contracts to be appealing in the first place. Greed causes speculators to want to control eight to ten times the volume for the same up-front investment. Greed also causes individuals to think, as the military is fond of saying, "in short time." Most Americans desire things now, not later. Almost every aspect of our culture reveals a passion for the here and now—low savings rates, high short-term consumer debt, billions of credit cards, new cars with seven-year payments rather than older cars, and house-rich/cash-poor two-wage-earner couples living in the suburbs. Is it any wonder why futures markets have exploded in terms of the volume of trades? They offer to the speculator a medium where greed is allowed to flourish. Leveraged contracts and liquidity (the here and now) are the norm in the futures markets.

The hallmark of successful speculators is that they understand the greed within themselves and others, and they use it to their advantage. They formulate trading strategies that limit their own greed and exploit other peoples' greed.

Fear, like greed, is a very powerful characteristic. At its worst it causes panic, and at its best it increases wisdom. Fear sometimes generates panic, which is usually a reaction to something greed has caused to happen whereby the event seems catastrophic. Some would say that the speculator begins to act irrationally. But panic, fear, and greed are very rational emotions—they only appear to be irrational. The fear/wisdom combination is a very useful characteristic in that it helps the speculator determine a market plan and stick to it. Fear, properly used, helps speculators know their limitations.

Puking

Some like to use the term "ego" rather than the crude term "puking" to describe the behavioral concept, but puking is more than just ego. Puking is the ability to recognize that you have made a mistake, correct it, and re-evaluate it, time after time. Long-term successful speculators have strings of unsuccessful trades, sometimes nine out of ten are not successful, but they correct them quickly and re-evaluate. That is to say, they puke—they recognize the mistake, get rid of it (puke), and re-evaluate. Successful traders have self-confidence but not the arrogance to assume or believe that they are always right. Pukers know that they will make many mistakes and have few successes. The mistakes don't drag them down—they know that with enough re-evaluation they will eventually win, and win enough to counterbalance their losses. There is an old saying that reinforces the concept of puking: Markets are never wrong, but people often are.

Unsuccessful traders believe they have found the truth, they alone have it, and markets will come around to their way of thinking. Successful speculators follow the logic that they think they know what the markets will do, but if they are wrong, then they will get out and try again. Unsuccessful traders believe in individual superiority, successful traders believe in market superiority.

Discipline

Ron Frost, former Vice President of Marketing at the Chicago Mercantile Exchange, is fond of saying, "Without discipline, the rest is meaningless." Ron outlined a speculative strategy based upon the letters of the word *discipline.*

*D*etermine what your profit objective is and what your loss limit is.
*I*nvestigate new trading techniques and tactics.
*S*et your priorities before making a trade.
*C*oncentrate on the market signals.
*I*ntelligently apply your rules of trading.
*P*lan your trade and trade your plan.
*L*earn to make decisions without always second guessing yourself.
*I*solate yourself from the crowd.
*N*ever look for excuses or anyone else to blame. Be the master of your own destiny.
*E*xecute your plan. Goals are dreams that are acted upon.

Discipline is a behavioral characteristic that, if utilized, will help speculators develop trading plans and follow the plans once executed. Discipline is the counterbalance for greed. Suppose that your trading plan is to buy only three gold contracts, but the market keeps moving up. Greed will encourage you to buy more, while discipline will maintain your trading plan.

Trading Plans

Success in speculation doesn't just happen. Oh sure, there are all kinds of stories and examples of the little old lady from Anytown, U.S.A., who got the "hot" newsflash from her cousin and got rich in cocoa futures. Or the used car salesman who just knew silver prices were going up and made a killing and then retired to an exotic island and lived happily ever after. No doubt some of the stories are true, or at least partially true, but for every little old lady or used car salesman story, there are probably one hundred other stories with not so happy endings.

Most speculators don't trade with a definite plan and most don't stick to their plan. But a plan is necessary for long-term success. Plans don't have to be complex, but they do have to be followed. Speculating with futures, as an investment, must involve planning and execution of plans.

Components of Trading

A trading plan needs a situation analysis, goal(s), and strategy(ies) as a minimum. Other components can be added, such as procedures or action plans, depending upon the speculator's preference.

Situation Analysis

A situation analysis is simply background research to determine what you have to work with and sufficient information about the commodity to formulate an intelligent strategy that will reach individual profit/risk goals. What you have to work with includes personal skills, liabilities, and goals as well as physical and financial assets. Information about the commodity is necessary; however, it can be sketchy or detailed depending upon the speculator. Obviously more information is preferred to less—up to a point. Sometimes so much information is gathered that it costs more than any potential profit or it can't possibly all be used, and indeed, may prove to be overwhelming and thus detrimental. A situation analysis involves the following as minimum:

1. Attitude towards risk
2. Ability to handle risk
3. Inventory of skills
4. Commodity information

The first step is to do a self-analysis of what you have to work with, such as your own abilities. Try to determine your attitude towards risk and your ability to handle risk. You may love to take risks, but can you handle the risk? Do you lay awake at night worrying about the trade you made yesterday? Will a three-day limit move against you put you in the hospital or the morgue? Remember that attitudes towards risk vary with the type and amount of dollars at risk. An individual might get a rush from risking one or even twenty dollars at bingo, on a lottery, or on a blackjack table, but the same individual might not handle the loss of income stability or the amount of funds at risk on a futures contract.

What skills do you possess to help you invest in commodity futures? Do you have knowledge of the particular commodity or industry? What level of knowledge do you have about price analysis?

A commodity inventory involves looking at what the commodity's price has done in the past and forecasts of what it will do in the future. Additionally other information about the commodity can be added, depending upon how much detail a trader desires, including such factors as market structure, participants, and pricing mechanisms. Past price information can be obtained by keeping your own records, getting historical price information from free sources such as universities, or buying information from private firms. Forecasts can also be obtained in much the same way. Price analysis is discussed in Chapter 14 for those who want to do their own forecasting or for help in understanding other people's forecasts.

Goals

A goal needs to be realistic and specific. There is nothing wrong with having an overall goal that is general, but specific goals are necessary to guide short-term actions. For example, you may have an overall goal of being a millionaire. However, specific goals are necessary to achieve the overall general goal. Simply having a goal of being a millionaire will not help much when speculating with futures contracts. Several specific goals involving speculation with futures contracts may contribute to the overall goal.

A specific goal may be a profit or price objective. A trader, after looking at the situation analysis, determines that Treasury bond futures are overbought and are likely to decrease at least $2. The specific goal is to capture the $2 decrease in price, should it occur. The goal is specific and realistic—the trader isn't trying to get his entire million dollar fortune from this one price move in T-bonds, only the expected $2.

Strategies

It is very important to follow through after a goal is set to determine how to achieve the goal. A procedure is a method to achieve a stated goal. A procedure for the goal of capturing the $2-00* down move in Treasury bonds might be: Issue a market-if-touched sell order when the price of March T-bond futures reaches $70-00 and when the price falls to $68-00, offset.

This is a very rigid procedure in that the goal will be achieved only if the price reaches $70-00 and then falls to $68-00. More flexibility can be achieved by developing a strategy. A strategy involves procedures and if-then criteria. The

---

* Treasury bonds and notes are quoted in 32nds. The convention is to quote the full dollar amount first followed by a dash and then the 32nds. A price of $98-16 is ninety-eight and 16/32 dollars per $100 face value of the bond.

rigid procedure just exemplified could be modified as follows: From current price levels of the March T-bond futures, if prices decrease more than 3/32, sell futures and protect with a stop placed at 1/32 above the sell price. Allow three place and lifts with the 3-1 criteria before re-evaluating. If a short position is placed and prices begin moving down, trail stops down with a 2/32 band. Position will be offset either by stops and/or when prices get to the $68-00 level. Build a pyramid with three initial contracts followed by two when prices get to the $69-10 to $69-20 range and a final sell of one contract when prices get to the $69-00 level.

Once a strategy is developed, it is important to stick to it. A well-thought-out strategy is the best way to control greed. Without a goal and strategy, a trader is tempted to let emotion, fear, and greed dominate his or her trades. A trading plan removes these behavioral problems and allows a trader to focus intelligence and skills through discipline.

A Word about a Simple Plan

Many successful speculators adhere to a very simple trading plan: Cut losses and let profits run. In other words, when your trade is losing money, offset fairly quickly and cut the size of the loss. When your trade is making money, ride the run for as long as possible. Don't get out too quickly. The plan operates on an old rule of thumb about trading in the futures market: About 85 percent of speculators lose money in the long run, and of the 15 percent who make money, they have losses 75 percent of the time. That is, speculators who make money in the long run with futures contracts have only one out of every four trades that is a gain. The one trade with a gain must have enough profit to cover losses on three other trades—thus the maxim of cutting losses and letting profits run.

There is an old business saying, "You can't go broke taking a profit" that has some validity in certain business settings. If a certain business deal offers a profit, no matter how small, it is generally better to take it and not be greedy. If enough low-profit deals are made, without having too many that have a loss, then a businessperson can make enough overall to have a thriving business. However, in futures trading this old saying may not be worth much and in fact may be a dangerous position to hold because most trades result in a loss. If a speculator has a string of losses, albeit small, and then has a trade that is profitable, but is quick to take the profit, then the profit will not be enough to cover the small numerous losses. Which means that in futures it is possible, and indeed highly probable, to go broke taking a profit.

Position Traders

With the exception of spreaders, all speculators are concerned with three things: (1) the level of prices, (2) the direction prices will go (trends), and (3) when the direction changes (turning points). All simple and complicated strategies and plans are built around these cornerstones of speculating: *Levels, Trends, Turnings (LTTs)*. If you memorize these three items, any strategy or plan you develop will naturally include them and be complete. Consider as an example the following situation:

A friend of yours works in the gold industry and has informed you that several old mine sites that have been abandoned have been purchased by a group of companies that has a new technology to extract gold from very-low-grade ore. He tells you that current cost of extraction for gold from typical ores runs about $350 per ounce, but with the new technology it should be no more than about $300. Your friend is a reliable source, and furthermore your research shows that indeed several old mining companies have been purchased by a new company that is a separate venture involving the current big players in the gold mining business. You look at the past price movement for gold and observe that it has maintained near $350 per ounce *(level)* with a slight **bear** trend over the last several weeks *(trend)*. When will it go up *(turn)*? In trying to answer the *LTTs*, you can start to develop a plan or strategy.

Your thinking might take the following path: If you feel that the new technology will be substantial, then gold prices should continue ***trending*** downward for the next several months and in fact change the overall ***level*** of price that gold will settle from around $350 per ounce to something between $300 and $350. The level probably won't be $300 because the new gold will not be a substantial portion of gold being mined, but it will not be insignificant, thus it should have some effect on reducing the overall level of gold prices. It is not likely that the price **trend** will turn up anytime soon because the supply of gold will increase without any major changes in the demand for gold. Your **plan** now can be formed: *The long-term level of gold prices will trend downward from $350 per ounce to at least $340 (a guess) with no major turns upward unless major changes occur in demand for gold.* **Strategy:** Sell five gold futures at the current price of $350 per ounce with a set stop at $355 per ounce. If the trend continues down, then when prices get to $345 per ounce, sell three more contracts and trail the set stops down to $2 per ounce above the initial selling price for each contract. If the price reaches $340 per ounce, liquidate all contracts and take the profit. If prices continue down, when they reach $338 per ounce, sell three contracts and set a stop at $339 per ounce. If the price trend continues down, sell two more contracts when it goes down another $2 per ounce, and then sell one more contract when it reaches below another $2 per ounce (with set stops set at $1 above initial selling price and trailed downward).

This strategy is very conservative. It builds a correct pyramid with trailing stops to protect profits. Because the new level is a guess, the strategy takes all profits out and rests at the guessed-at level of $340 per ounce. If the trend continues down, then a new smaller pyramid is developed to capture some of the profits as the price seeks a new uncertain level. If the whole plan is wrong and prices do not go down and it fact turn up, then the stops get the contracts liquidated with only a small loss.

Notice that any plan that a speculator develops has to include thinking about the LTTs. Even a simple plan includes LTTs whether or not the speculator consciously thinks about them. If a speculator thinks that gold prices will increase and then very quickly buys two contracts, without a plan or strategy, then de facto LTTs were used. The speculator believes that from the current level of prices, a bullish trend will occur and will not turn around and go down before he can sell the contracts and make money. The point is, all speculators either consciously or unconsciously consider LTTs. A more powerful strategy can be developed if LTTs are consciously considered. Forecasting tools must be used to estimate the LTTs. Chapter 14 covers the basic forecasting tools used in the futures and options markets.

Managed Trading Systems

It is possible to speculate with commodity futures and not be knowledgeable about the pros and cons. There are numerous firms that will trade for you with a device called the managed account or system. The idea is very similar to mutual funds. A managed account allows individuals to put a fixed amount into a pool. The pool is then invested by an account executive or manager in several different futures contracts. The pool is often treated as a portfolio in that several different futures contracts are traded such that if one is in a losing position, there will be, hopefully, others that are in a profit position.

Types of Systems

There are as many different types as there are accounts. However, they generally fall into a few categories.

Fixed Investment/Margin Calls. This type of account requires a set amount to be in the pool—say $10,000—and the whole amount is invested. Any margin calls for additional funds are passed proportionally to each investor. This type of account is also called *open ended.*

Fixed Investment/No Margin Calls. A fixed amount is required, but the whole amount is not invested. The portion not invested is reserved for margin calls (also called a *closed ended* account). If this reserve is not enough, some positions will be liquidated to make the required margin calls. This is obviously more popular than a system that requires additional funds for margin calls, because the amount invested remains constant for some fixed period.

The open and closed ended accounts constitute most managed commodity accounts. However, there are probably as many other types as human imagination can devise. It is wise to get a copy of the agreement from the firm offering the account and have a competent attorney review it and explain your liability and the general terms of the agreement. Almost all major managed accounts are partnerships with a general partner who is usually someone at the firm that issues the fund. Managed accounts earn money by charging a sales fee for each unit sold, a commission based upon the size of the fund, and/or both. Rarely do managed funds collect fees from the profits earned as their sole source of revenues.

Most managed accounts are sold in units of $1,000. If you purchase a unit for $1,000 and are credited with only $975, then you have been charged a sales fee up front. You may also be charged a commission on a regular basis as long as you hold the unit.

Another type of managed account has emerged in the last few years that guarantees the investor that the principle will be returned in a fixed period (usually two to three years). If the investor puts up $10,000, the manager will guarantee that the investor will get his $10,000 back in three years as a minimum. If the account makes money, then the original investment plus profits will be returned. The manager of the account can make such a guarantee because he does one of two things: (1) discount back to present value what $10,000 in three years is worth today (at ten percent, it is worth $7,513) and put that amount in a three-year Certificates of Deposit that will mature worth $10,000, or (2) buy a zero coupon three-year bond. The manager invests the difference ($10,000 – $7,513 = $2,487) in futures accounts and uses the CDs as margin.

This is a very risk averse strategy, but likewise is very appealing to investors who want to be assured they will at least get their original investment back at some point. The individual investor must realize that she has an opportunity cost for the $10,000 invested for three years. The guarantee is for the original investment only, not the original investment plus interest.

The performance of managed funds is erratic. Most funds are not consistent from period to period, especially in generating profits. Some are very consistent in generating losses. It seems that managed accounts' performance mirrors the general trading of individuals—lots of losses and few winners. The performance of managed accounts (funds) is regularly reported in publications such as *Financial World* and *Futures*.

PROBLEMS

1. Holly believes that the price of corn will go down during the next two months by at least 20 percent from its current level of approximately $3.40 per bushel (about $.68 to $.70 per bushel decrease). She thinks that before it starts to decline, it will actually increase to the $3.50 per bushel level. Holly's strategy is as follows: Have a market-if-touched order in for five contracts to sell at a price of $3.50 per bushel with a set stop order at $3.55 per bushel should the contracts get sold. If the price moves to $3.45 per bushel and for each five-cent-per-bushel move down thereafter, trail the stop down to within $.03 of the market. When the market decreases to $3.30 per bushel, sell four more contracts with the three-cent set stop order in place. For each ten-cent price move down thereafter, sell (n – 1) contracts (where *n* represents the previous number of contracts sold). For example, if the price moves to $3.20 per bushel, then the number of additional contracts sold would be 3 (4 – 1). When the price reaches $2.75 per bushel, liquidate all contracts.

   A. What are Holly's assumptions about the LTTs with this strategy?

   B. Calculate Holly's gross returns (ignore brokerage fees, margin costs, and the like) under three price movements: (1) The price moves down $.01 per bushel per day from $3.52 per bushel to $2.70 per bushel; (2) the price moves down $.01 per bushel per day from $3.52 per bushel to $3.32 per bushel and then moves up $.01 per bushel per day till it reaches $3.60 per bushel; (3) the price moves up $.01 per bushel per day from $3.40 per bushel to $3.90 per bushel.

2. Harry is very scared of the futures markets. He speculated with futures contracts for one three-month period and ended up losing almost $2,000. Harry says he will never trade again. His three-month trading spree went as follows: Harry received a $4,000 check from the estate of a cousin and a few months later got an income tax refund of $4,000 because of overpayments and some casualty losses. His friend was a broker and suggested that he should put the money where it could potentially earn a higher return than passbook savings at a bank with not much more risk. Harry basically lives paycheck to paycheck with only about $2,000 in the bank in cash and savings. He tells his broker friend that he is very risk averse and doesn't want to lose the $8,000. His broker friend shows him the gold price chart for the past few months and points out that gold prices have remained fairly stable between $400 per ounce and $420 per ounce. The broker tells Harry that all the indications show that gold is ready for a major bull move. He suggests to Harry that he buy one gold futures contract ($5,000 margin) at the current market price and place a stop loss order $2 per ounce below the buy price. The broker points out that if gold prices don't go up and in fact go down, then Harry will be stopped out with about a $2 per ounce loss, or $200 on a 100 ounce contract. The broker points out that if prices go up as he anticipates, then Harry's $5,000 investment will earn a very nice return in a short time.

   Harry begins to trade. Neither he nor his broker changes his viewpoint that the market will go up. The market during the three months of trading never leaves the band of prices that it has been trading at for the past several months (i.e., between $400 and $420 per ounce). How did Harry lose $2,000?

Investing as a Hedger—The Basics

"Don't try to buy at the bottom and sell at the top. This can't be done—except by liars."

—Bernard M. Baruch

Futures markets provide the investment opportunity to protect profit levels, margins, portfolio values, capital gains, inventory values, and a whole host of other economic variables by hedging. Hedging is simply shifting the risk of price change in the cash market to the futures market. The process of shifting risk involves simultaneously taking an opposite position in the cash market relative to the futures position. Put another way, hedging is replacing a future cash transaction with a futures contract. Consider the following example as an illustration.

A gold dealer's primary business is to buy gold with the intent of selling it sometime later, hopefully for more than he paid for the gold. He may be able to find a buyer in a few minutes, a day, or perhaps it will take a week or more to structure the proper deal for sale. Between buying and selling, the gold dealer is exposed to downside price risk because ownership of the gold has transferred to him during the period between buying and selling. He can lay off or shift that downside price risk to the futures market by hedging.

If the dealer bought 100 ounces of gold bullion for $400 per ounce, he is said to be long the cash gold, so an opposite position in the futures market is to sell futures contracts. Also, because he intends to sell the cash gold sometime in the future, the hedge would consist of selling a gold contract in the futures market.

There are three primary rules or tests for a proper hedge: (1) The investor must hold opposite initial cash and futures positions; (2) final cash and initial futures positions must be the same; and (3) if cash price risk is declining prices, then enter a **short hedge** by selling futures, and if cash price risk is increasing prices, then enter a **long hedge** by buying futures. The three tests are useful for hedgers to determine if they are properly hedged. Any one test is necessary but may not always be sufficient, thus the need for two more. With some hedging situations it is difficult to determine either the initial cash position or the final cash position or even the price risk, but one of the three tests will always confirm whether the **hedge** was properly placed.

The example in Table 6.1 shows the results of short hedge under both price increase and price decrease scenarios. The gold dealer wants downside price protection; when prices do move down, she is protected through a short hedge. She has shifted the downside price risk to the futures market by selling a futures contract. If the price moves down, she can buy the futures contract back at a lower price and thus offset the declining price of the cash position. However, if the price moves up, the cash position gains in value but the futures loses in value. The hedger is, therefore, protected for both downside and upside price moves. This is a feature of hedging that always seems to cause anger from new hedgers: The gains they believe should be theirs when cash prices increase are lost through the futures market. Hedgers are truly protected from both upside and downside price movements even when they wish they weren't.

Table 6.1 Gold Dealer Hedging Example

Date	Cash Transactions	Futures Transactions
November 1	Buys 100 ounces of gold bullion for $400 per ounce.	Sells 1 December gold futures at $410 per ounce.
	Price Decrease	
November 5	Sells 100 ounces of gold bullion for $390 per ounce.	Buys 1 December gold futures at $400 per ounce.
	$10 per ounce loss on cash transaction.	$10 per ounce gain on futures transaction.
	Net Hedged Price = $390 + $10 = $400/ounce	
	Price Increase	
November 15	Sells 100 ounces of gold bullion for $410 per ounce.	Buys 1 December gold futures at $420 per ounce.
	$10 per ounce gain on cash transaction.	$10 per ounce loss on futures transaction.
	Net Hedged Price = $410 – $10 = $400/ounce	

Basis and Basis Changes/Net Hedge Price

Net hedge price is the price actually paid or received in the cash market plus or minus the gain or loss on the futures transaction. A **net hedge price** is therefore a net buying price or a net selling price. What a hedger really wants to know is: What is my net position considering both my cash and futures transactions? Net hedged selling prices and net hedged buying prices are discussed in greater detail in a later section.

Notice that the relationship between the cash and futures price in the gold dealer example remained the same. The difference between the cash gold price on November 1 and the December gold futures was $10 per ounce as it was on November 15 when the gold dealer sold the gold and lifted her hedge. This numeric difference between the cash price of a commodity or financial and a futures contract is called **basis**. The cash market for a commodity or financial instrument is not the same market as futures markets for the same entity and does not respond in exactly the same way at the same moment to new information, thus the two prices will generally be different as shown in the example of Table 6.2. Moreover, basis (a $10 difference) does not always stay the same. A discussion of why there is basis and what it is composed of is included in each chapter on hedging different futures contracts. Suffice it to say at this point that there is basis and it changes. Basis is simply the difference between the cash and futures prices and, therefore, is subject to change.

What hedging does is shift the risk of absolute or total price movements to basis movements. The risk of the change of absolute cash prices is substituted for the risk of basis changes. In other words, with hedging you can protect against the cash price moving against you, but you cannot protect against the relative movements of the cash and futures prices. Thus, an element of risk still exists with all hedges—speculating on the basis changes. That is why there is no such thing as "locking in a price" by hedging. You cannot protect against basis changes so that no locking of price exists.

Let us go back to the gold dealer hedging example and show the effects of basis changes in Table 6.2. In the first panel, the futures price fell more than the cash price did, thus the basis changed from $10 per ounce to $8 per ounce.

Table 6.2 Gold Dealer Hedging Example with Basis Changes

Date	Cash Transactions	Futures Transactions	Basis
November 1	Buys 100 ounces of gold bullion for $400 per ounce.	Sells 1 December gold futures at $410 per ounce.	$10 per ounce
	Basis Improvement		
November 5	Sells 100 ounces of gold bullion for $390 per ounce.	Buys 1 December gold futures at $398 per ounce.	$8 per ounce
	$10 per ounce loss on cash transaction.	$12 per ounce gain on futures transaction.	$2 per ounce
	Net Hedged Price = $390 + $12 = $402/ounce		
	Basis Deterioration		
November 15	Sells 100 ounces of gold bullion for $390 per ounce.	Buys 1 December gold futures at $402 per ounce.	$12 per ounce
	$10 per ounce loss on cash transaction.	$8 per ounce gain on futures transaction.	$2 per ounce
	Net Hedged Price = $390 + 8 = $398/ounce		

This reduction in basis is called a narrowing of the basis. When the basis narrowed from the time the hedge was placed until it was lifted, the amount of the basis change was an increase in the net price for the hedge of $2 per ounce. In the second case, the futures price fell less than did the cash price, and basis changed from $10 to $12. When the basis widened, it deteriorated the net price by $2 per ounce. The gold dealer's hedge protected her against the absolute change in the cash price of gold (a decline of $10 per ounce), but it did not protect her from the relative change in the price of the two markets (a net decline of $2 per ounce). The relative change can be an improvement or a deterioration, but it is the risk of hedging that cannot be offset. Proper planning can help estimate the probability of basis changes and when they are more likely to occur, but it cannot remove the risk of basis change.

Types of Hedges

There are two basic hedges: a short hedge and a long hedge. A short hedge is used to protect against declining values of the cash position (declining prices); while a long hedge is used to protect against increasing values of the cash position (increasing prices). The previous example with the gold dealer was a short hedge. The dealer was protecting against the declining value of the cash position.

A long hedge would involve the need to protect against the risk of the cash position increasing in value. Consider the example of a jewelry manufacturer who has a chance to sell gold jewelry today for delivery in one month. He does not currently have the gold in inventory to make the jewelry, nor does he want to buy it today and hold it for three weeks before he starts to actually manufacture the jewelry. However, he must price the jewelry today in order to make the sale. Generally the manufacturer will price the jewelry today based on today's cash gold price, but then now faces the risk that gold prices will increase between today and the time he actually buys the gold to use in making the jewelry. By using the futures market, he could hedge that risk by placing a long hedge, as shown in Table 6.3.

Table 6.3 Jewelry Manufacturer Long Hedge with Basis Improvement

Date	Cash Transactions	Futures Transactions	Basis
October 1	Sells the equivalent of 100 ounces of gold in jewelry based on a gold price of $400 per ounce.	Buys 1 December gold futures at $410 per ounce.	$10 per ounce
October 21	Buys gold to begin jewelry manufacturing at $410 per ounce.	Sells 1 December gold futures at $422 per ounce.	$12 per ounce
	$10 per ounce loss on cash transaction.	$12 per ounce gain on futures transaction.	$2 per ounce
	Net Hedged Price = $398/ounce		

The jewelry manufacturer was protected against the $10 per ounce increase in the cash gold prices by the $12 increase in the futures prices. Thus, he actually paid $398 for the cash gold, which was $2 less than the estimated price of $400 per ounce because the basis improved by $2 per ounce. Had the basis deteriorated, then his cash price would be $2 more per ounce than $400, the amount of the basis change as shown in Table 6.4.

Table 6.4 Jewelry Manufacturer Long Hedge with Basis Deterioration

Date	Cash Transactions	Futures Transactions	Basis
October 1	Sells the equivalent of 100 ounces of gold in jewelry based on a gold price of $400 per ounce.	Buys 1 December gold futures at $410 per ounce.	$10 per ounce
October 21	Buys gold to begin jewelry manufacturing at $410 per ounce.	Sells 1 December gold futures at $418 per ounce.	$8 per ounce
	$10 per ounce loss on cash transaction.	$8 per ounce gain on futures transaction.	$2 per ounce
		Net Hedged Price = $402/ounce	

Computing Net Hedge Prices

Computing a net hedged price is simple mechanically but can be confusing. It is important to keep straight that a short hedge indicates the need to compute a net hedged selling price because the primary action is to sell the commodity, whereas a long hedge indicates the need to compute a net hedged buying price because the primary reason to enter a long hedge was to buy the commodity.

Net Hedged Selling Price

For short hedges the calculation to compute a net hedged selling price is to take the actual cash selling price and add futures profits or subtract futures losses; thus:

$$\text{NHSP} = \text{CP} + \text{FP}$$

where

NHSP = Net Hedge Selling Price

CP = Cash Selling Price

FP = Profit or Loss on Futures

In Table 6.2 the gold dealer sold her gold bullion for $390 per ounce and had a gain on futures transactions of $12 per ounce. Therefore,

$$\begin{aligned} NHSP &= \$390 + \$12 \\ &= \$402 \end{aligned}$$

Another method of calculation is to take the cash price at the time the hedge was placed (buying price) and add or subtract the change in the basis depending on whether or not it was an improvement (add) or a deterioration (subtract). This is expressed as:

$$NHSP = ICP \pm BC$$

where

ICP = Initial Cash Price (Buying)

BC = Basis Change

In the same example in Table 6.2,

$$\begin{aligned} NHSP &= \$400 + \$2 \\ &= \$402 \end{aligned}$$

The $2 basis change was an addition because it was a basis improvement.

Net Hedge Buying Price

The formulas are similar for computing net hedge buying prices and net hedge selling prices. The application is different, however. With the NHSP when there was a futures gain, it was added to the cash selling price. For net hedge buying prices a gain in the futures is subtracted. The formula is

$$NHBP = CP - FP$$

where

NHBP = Net Hedge Buying Price

CP = Cash Buying Price

FP = Profit or Loss on Futures

In Table 6.3

$$\begin{aligned} NHBP &= \$410 - \$12 \\ &= \$398 \end{aligned}$$

There was a $12 gain on the futures transaction, which had the effect of lowering the cash buying price. Thus what the jewelry manufacturer really paid net for the gold was $398. The same logic applies to the second formula using basis changes. If the basis change is an improvement, then it is subtracted from the initial selling price. Again, using the example in Table 6.3, the results are

$$\text{NHBP} = \$400 - \$2$$
$$= \$398$$

A deterioration in the basis would be an addition.

In Table 6.4 the example would yield NHBPs using the two formulas as follows:

$$\text{NHBP} = \$410 - \$8$$
$$= \$402$$

or

$$\text{NHBP} = \$400 + \$2$$
$$= \$402$$

Net hedge prices are calculated using these formulas regardless of commodity (corn, cattle, T-bills, etc.). These formulas must be modified when different quantities of cash and futures are used. If a hedger had a 200-ounce cash gold position and hedged only 100 ounces (one contract), then the net hedge price formulas must be modified to reflect these mismatches of quantities. This can be mathematically expressed either on the cash or futures position. The easiest is to take the percent the futures position is of the cash (such as 50 percent) and multiply the futures gain or loss by this fraction, as

$$\text{NHSP} = \text{CP} + \text{FP}\ (.5)$$

This adjusted NHSP can now be multiplied by the actual cash quantity sold to yield the true net hedge returns.

Hedging Issues

Over and Under Hedging

Because futures contracts are standardized, especially with respect to the size of the contracts, it is often the case where the cash position needing to be hedged does not exactly match the quantity of a futures contract. Obviously, a trader with the need to hedge 200 ounces of gold could buy or sell two 100-ounce contracts, but a trader needing to hedge 150 ounces cannot enter 1½ contracts. Thus a decision must be made whether to over or under hedge, that is, to have a futures position larger than the cash position or smaller. Close positions do not need serious consideration such as a cash position of 90 ounces of gold and gold futures standardized at 100 ounces.

More difficult decisions involve a cash position of 30,000 pounds of feeder cattle and a standardized futures of 50,000 pounds. The decision is whether to not hedge or over hedge. Or consider a 60,000-pound cash position where the decision is to under hedge with one futures or over hedge with two. In the latter example, the trader is hedged for 50,000 pounds and has a 10,000-pound speculative cash position or hedged for 100,000 pounds with 40,000 pounds in a speculative futures position.

This position is not of real consequence to large block traders. Someone with a $2,050,000 cash T-bond position is not real concerned about being overhedged $50,000 or underhedged $50,000 (T-bond futures contracts are standardized at $100,000). But the decision is critical to small cash positions.

Unfortunately, there is no easy answer. The answer must be derived empirically for each commodity. Even then the answers are based on past information and thus are subject to being incorrect for the present. Even the methodology used to empirically test each commodity is not universally accepted.

Contract Month

Hedgers must also decide which contract month is the appropriate month. The general rule is to pick a contract month that is as close to the actual cash position as possible, but not before. If you plan to sell your T-bonds in November, then the December contract is the closest to the actual cash date, but not before. Picking the September contract would have the futures position liquidated in September and the cash position unhedged from September until November.

Two possible positions make the general rule inappropriate. First, sometimes the date the cash position will be liquidated or acquired is not always known. Therefore, the hedger must make a calculated guess. If the futures contract nears expiration and the cash position is still uncertain, then the hedger should liquidate the hedge and either remain unhedged or rehedge with a more-distant futures contract (called a **rolling hedge**). Secondly, if the anticipated cash liquidation or acquisition date coincides with a futures delivery month, care must be taken in selecting the futures contract month with which to hedge. Futures delivery months can be volatile, especially towards the last few days of trading when price limits are removed. The last trading day for most futures contracts falls roughly on the 20th of the month, thus if the cash position will be liquidated or acquired towards the end of the month, the next available futures contract should be used.

Furthermore, the general rule may not be applicable in certain situations. Some distant months may prove to be better hedging months than ones closer to the actual cash liquidation or acquisition. Sophisticated hedging programs explore empirically which futures month offers the best hedging results.

Rolling the Hedge

A rolling hedge is often used when uncertainty regarding the cash position's liquidation date exists. The hedger maintains a cash position or anticipated cash position and has it hedged with a nearby futures contract. When the futures nears expiration, the hedger will lift the initial hedge (offset the futures contract) and replace it with another **nearby** futures contract. This type of hedge is very useful when the length of time the cash position is held is uncertain but needs to be maintained with liquidity. This is illustrated with an example in Table 6.5. The gold dealer buys gold bullion and then hedges it with the nearby December contract. When December 15 arrives, the dealer still maintains the cash position, lifts the hedge, and rolls into another with a March contract. As a short hedge this strategy works well because the basis has a tendency to narrow (in this example by $28/oz.) as a futures contract approaches expiration.

Table 6.5 Example of a Rolling Hedge

Cash Position	Futures Position	Basis
November 1		
Buys gold bullion at $400/oz.	Sells December at $420/oz.	$20
December 15		
Maintains cash position	Buys December at $410/oz.	$12
(cash price at $398/oz.)	+ $10/oz.	change of $8
	Sells March at $438/oz.	$40
February 15		
Sells gold bullion at $378/oz.	Buys March at $398/oz.	$20
– $22/oz.	+ $40/oz.	change of $20
Net Hedge Selling Price = $378 + $10 + $40 = $428/oz.		

Target Prices

One of the most useful tools hedgers can use is target pricing. A target price is an estimate of the net hedge price. It is calculated by taking the current futures price and adjusting it with an estimate of the ending basis. The actual formula depends upon how the basis is calculated. Basis is calculated either by taking the futures price and subtracting the cash price or vice versa. Both methods are used; thus to properly use basis, it is necessary to find out how it was calculated. If basis is calculated by the formula

$$FP - CP = B$$

where

FP = Futures Price

CP = Cash Price

B = Basis

then the target price formula is

$$TP = CFP - EEB$$

where

TP = Target Price

CFP = Current Futures Price

EEB = Estimate of Ending Basis

If, however, the basis is calculated

$$CP - FP = B$$

then the target price formula is

$$TP = CFP + EEB$$

Estimates of ending basis values are usually developed by averaging several years of historical values. Sometimes, instead of an average, either the mode or median value is used. Most state land-grant universities have calculated basis values for the major agricultural commodities; however, basis values for other futures are very limited. Therefore, individual basis tables must be calculated for these futures contracts.

Consider as an example a gold dealer who is considering purchasing some gold bullion. She needs an estimate of what she will be able to sell the gold for later—a target price. She observes that the current quote on the nearby gold futures (March contract) is at $456 per ounce. She estimates that if she buys the gold today, she will sell it in two weeks (about February 15). She looks up her historical basis tables and finds that on average for the last seven years the difference between the March gold contract and the cash market on February 15 has been $12 per ounce (calculated FP – CP = B). That is, on average for the last seven years the March gold contract has been $12 per ounce higher than the cash market on February 15. Her target price (estimate of net hedge price) is $456 – $12 = $444. This means that if she can place the hedge at $456 and the basis turns out to be $12, she will receive a net hedge price of $444 on February 15. She can now decide what price she is willing to pay for gold today.

Double Whammy

One of the major mistakes novices make in hedging is called the **double whammy**. A double whammy is a loss in the cash and a loss in the futures. It is best explained with an example. Consider a gold dealer who buys gold and hedges it with a March gold contract. The dealer is worried that prices will fall while he holds the cash gold. The short hedge will protect that position. However, after hedging, the price starts to increase. The dealer starts to gain in the cash, but has losses in the futures and has to make margin calls. The dealer then decides that the hedge is no longer working and is certainly no longer necessary and liquidates the hedge, at a loss in the futures markets. He maintains his cash position because it is increasing in value. However, the price increase is brief and then starts a rapid decrease. The dealer rides the price down and sells the cash at a loss, which magnifies his earlier losses on the futures—that is, a double whammy. Had the dealer maintained the hedge, the margin calls would have stopped when prices fell and profits would have accrued on the futures side to offset the cash losses and the hedge would have protected the cash position. Table 6.6 shows the double whammy with a numerical example.

Table 6.6 Example of a Double Whammy

Cash Position		Futures Position
February 15		
Buys gold bullion at $400/oz.		Sells March at $420/oz.
	Prices Increase	
February 20		Buys March at $440/oz.
Cash at $420/oz.		– $20
	Prices Decrease	
February 25		
Sells at $395		
– $5/oz.		– $20/oz.
	Net Selling Price = $395 – $20 = $375/oz.	

PROBLEMS

1. Inga has placed a short hedge with silver futures with a basis of $.50 per ounce. What basis would she want to have when she lifts the hedge—$.40 per ounce or $.60 per ounce? Explain.

2. Boris owns 10,000 bushels of corn currently in storage. He has it hedged with two December corn futures contracts at $3.50 per bushel. When Boris put the grain in storage and hedged it, the cash corn price was $3.20 per bushel. Since he put the grain in storage, the cash price has risen to $3.60 per bushel and the December futures price has increased to $3.85 per bushel. Boris has received several margin calls, and when the broker called for the last one, Boris told her to liquidate his futures contracts because he no longer needed the hedge. The broker liquidated the futures contracts at $3.80 per bushel. Boris still has the grain in storage. Boris has vowed that he will never hedge again because his hedge did not work. Comment.

3. Shontell is considering planting wheat on her farm. She checks with the local elevator and finds that the average basis on July wheat on the Kansas City Board of Trade in the first week of June is on average $.20 per bushel (with a low of $.10 and a high of $.32). She calls her broker and finds out that the current quote on July KC wheat is $4.00 per bushel. Her farm foreman tells her that the average cost of production is $3.90 per bushel. Her farm foreman tells her that it's time to plant wheat, he has always planted wheat on the farm, and he needs her decision by the end of the day. What is Shontell's profit potential with wheat on a bushel basis?

4. Short hedgers want the basis to move a certain direction to have a basis improvement, likewise for long hedgers. What would be a good general rule of thumb for both short and long hedgers to use?

CHAPTER 7

Hedging Strategies: The Grain and Oilseed Complex

"How little you know about the age you live in if you think that honey is sweeter than cash in hand."

—Ovid

The first modern futures contracts were developed for the grain markets, and thus the grain complex has a long and detailed futures market history. The grain industry is where most futures traders developed their skills. They have, however, lost their lock on dominance in the futures trade in recent years. The financial futures have far outpaced all the agricultural commodities. By the mid-1980s they had grown to be the dominating contracts, but grain futures contracts remain strong and viable. There is still a strong economic need among hedgers to control the risk of cash grain price movements especially considering the fact that the grain industry is strongly influenced by world markets. The United States continues to lose market shares in the world grain markets. Developing and developed nations produce more grain and oilseed crops for domestic food security and as a means to gain foreign exchange. Each passing year moves the grain trade into world economic forces with less and less influence by a single country. Therefore, more volatility exists now in the grain and oilseed markets and will continue to grow in the future. Hedging becomes even more important for producers, shippers, and processors, and more opportunities exist for speculators.

Futures trading in the grain and oilseed complex first began on the Chicago Board of Trade in wheat, corn, and oats in 1865. During the mid-1800s, hard-surfaced roads were not common between farms and markets anywhere, including the Midwest where a major share of U.S. grain production was then and still is concentrated. This lack of roads, coupled with inadequate storage, caused major marketing problems and wide seasonal price swings for a commodity that is produced in only one cycle each year. Futures trading offered an orderly marketing alternative for producers, warehouses, and grain processors alike.

Cotton futures trading began in New York in 1870. The Kansas City Board of Trade, the Mid-America Commodity Exchange, and the Minneapolis Grain Exchange followed the Chicago Board of Trade's lead with wheat futures contracts in 1877, 1880, and 1885, respectively, and the Midam opened a corn contract in 1880 as well. Soybean trading was added much later, in 1936, as were contracts for the soybean complex (oil in 1950 and meal in 1951). Other grains and crop contracts have been attempted, including grain sorghum, rye, and apples.

Other current but lower-volume contracts include rice, flax, seed potatoes, barley, and canola. Each of the grain contracts is traded in 5,000-bushel units (roughly the equivalent of one rail car), except the Canadian contracts (Winnepeg flaxseed, barley, canola, and wheat), which is traded in 20-metric-ton units (approximately 44,000 pounds or a semitruck load), and the Mid-American mini-contracts (typically 1,000 bushels).

Two broad categories of hedgers are discussed in this chapter—producers and merchandisers. The intent of the chapter is to introduce and illustrate some of the more traditional ways that hedging in the grains and oilseeds is accomplished. The concepts presented are by no means an exhaustive list. Individual creative efforts makes the list infinite.

Producers

Producers of grains and oilseeds have the risk of price changes while the crops are growing and to a great extent while the crops are in storage prior to sale. Thus, the two major types of hedges for producers are production hedges (also called anticipatory hedges) and storage hedges (also called carrying hedges or inventory hedges).

Production (Anticipatory) Hedges

A production hedge is a hedge placed before or during the growing period for the crop. It is called an anticipatory hedge because the producer anticipates producing and/or harvesting the crop and selling it at some point in the future. Production hedges are short hedges. The producer's price risk is that during the growing period the cash price will decrease for two reasons. First, the producer made his decisions on what and how much of each commodity to grow based on expected prices. If prices fall, especially if they fall below a break-even level, those decisions will later appear to be erroneous. Secondly, agricultural producers (and all other producers as well) view a decline in price as a loss even if they were unable to sell a crop at that price because it had not yet been harvested. Table 7.1 shows the basic concept of a corn producer who places the hedge a few days after planting on May 15 and plans to sell the cash grain at harvest on November 15. No basis change is assumed for the moment. If prices fall as they normally do from the spring to the fall, then the hedged producer received a net hedge price of \$3.50 rather than the unhedged price of \$2.90. If cash prices increase, the hedged producer still received \$3.50 rather than the unhedged price of \$4.00.

Table 7.1 Corn Production Hedge

Cash Position	Futures Position	Basis
May 15		
Crop planted, local cash price \$3.50/bu.	Sell December contract at \$3.70/bu.	.20
	Price Decrease	
November 15		
Harvest and sell on local market for \$2.90/bu.	Buy December contract at \$3.10/bu.	.20
	+ .60/bu.	no change
Net Hedge Price = \$2.90 + .60 = \$3.50/bu.		
	Price Increase	
November 15		
Harvest and sell on local market for \$4.00/bu.	Buy December contract at \$4.20/bu.	.20
	– .50/bu.	no change
Net Hedge Price = \$4.00 – .50 = \$3.50/bu.		

Producers must have a stated goal and stick to the goal because it is tempting to lift the hedge when prices start up. If the producer was satisfied that $3.50 per bushel plus or minus basis changes would cover production costs and an adequate profit, then the producer really doesn't care whether prices increase or decrease if hedged properly. Further, this dissatisfaction with futures losses as cash prices move upward is one of the least understood and most problematic elements of hedging with futures contracts.

Storage Hedges

A producer may decide not to sell at harvest but rather to store the grain commercially or on-farm for later sale or for later use as feed. The price risk that a producer faces while the grain is in storage is the same as during production. A decrease in the cash price during storage is destructive because storage costs accumulate solely with the passage of time. However, from harvest time during the fall until late spring and early summer, the seasonal average price movement is up for most grains and oilseeds (other than for spring/summer harvested crops such as winter wheat).

The producer can short hedge to protect against decreasing prices, but most storage hedges involve capturing storage income through movements in the basis. The basis pattern for most storable commodities such as grains and oilseeds shows a narrowing trend towards maturity of the contract (see Figure 7.1). A storage hedge can capture the narrowing of the basis and provide protection against decreasing prices and at the same time provide storage revenues. Table 7.2 has a illustration of a storage hedge for corn placed in December and lifted in May. The storage hedge earned a $.55 per bushel increase in revenue over selling at harvest rather than storing until May.

Table 7.2 Storage Hedge for Corn

Cash Position	Futures Position	Basis
December 1		
Put corn in storage (cash price at $2.80/bu.)	Sell May contract at $3.40/bu.	.60
May 1		
Sell out of storage for $3.20/bu.	Buy May contract at $3.25/bu.	.05
	+ .15/bu.	change of .55
Net Hedge Selling Price = $3.20 + .15 = $3.35 bu.		

Producers can use storage hedges to justify and pay for the construction of on-farm storage. Properly placed storage hedges may provide storage income through basis movements to justify building on-farm storage. However, all direct

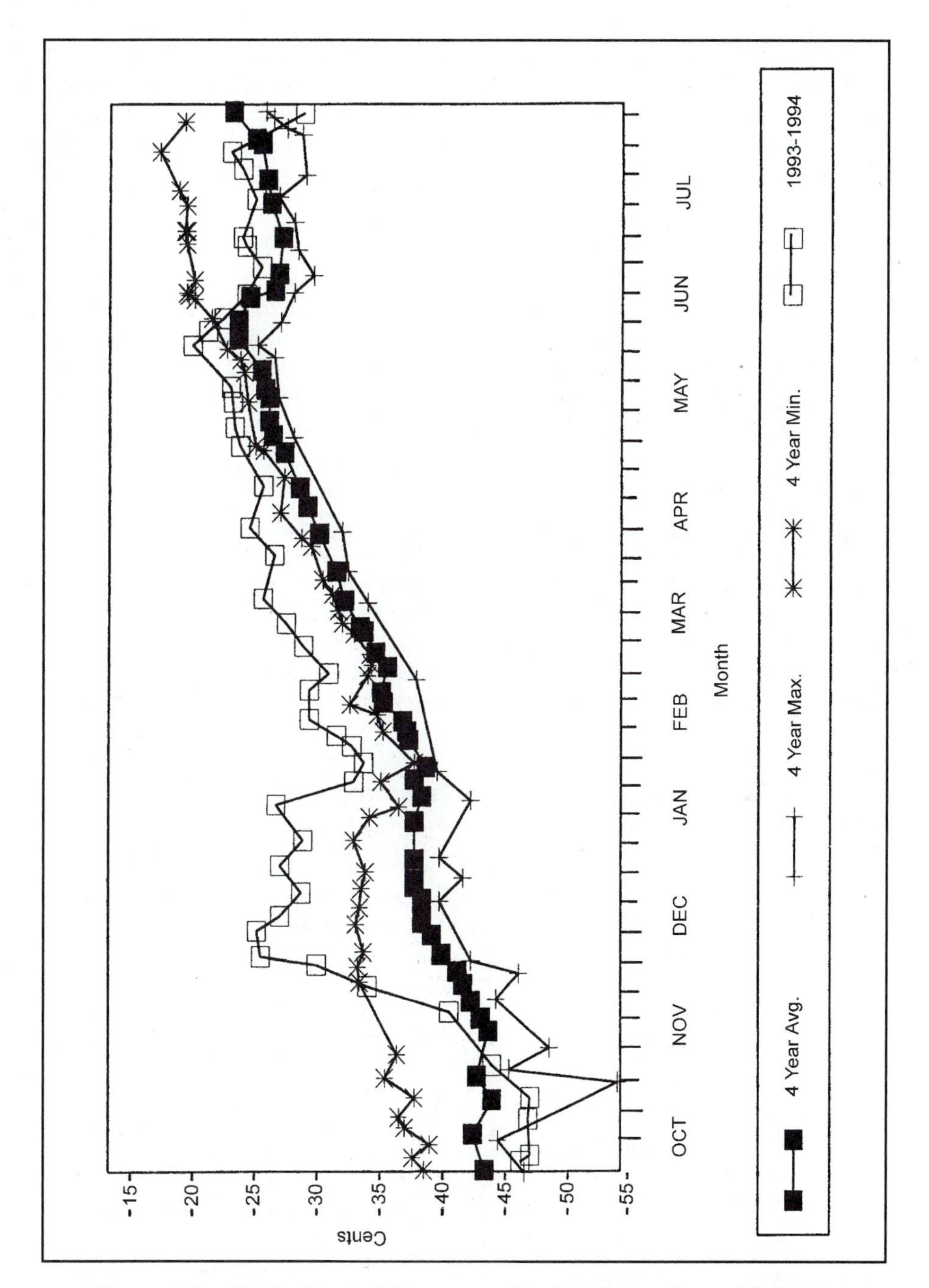

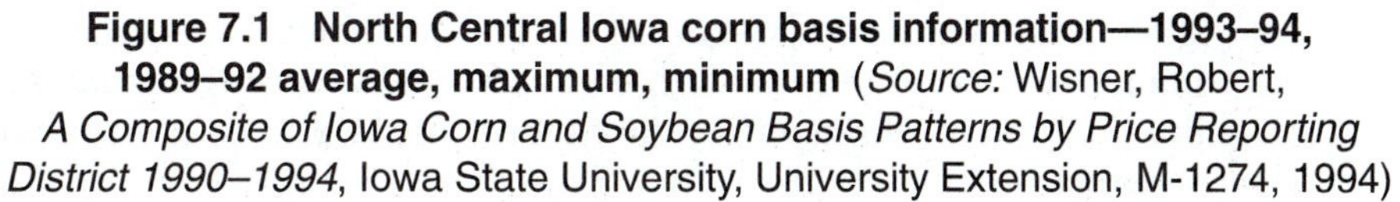
Figure 7.1 North Central Iowa corn basis information—1993–94, 1989–92 average, maximum, minimum (*Source:* Wisner, Robert, *A Composite of Iowa Corn and Soybean Basis Patterns by Price Reporting District 1990–1994*, Iowa State University, University Extension, M-1274, 1994)

and opportunity costs must be considered in the overall evaluation, as the cost of funds involved in storing grains is often sufficient to make storage unprofitable, that is, the basis does not increase enough to cover all costs. The cost of storing commercially needs to be less than the earned storage income from the hedge. This may or may not always be possible. Target prices can be helpful in evaluating the decision.

Rolling the Hedge

Producers who have production hedges but decide not to sell at harvest and instead store the grain or oilseed can roll their production hedge into a storage hedge. Table 7.3 shows an example of this process. The producer simply offsets the production hedge and replaces it with a futures contract that represents the likely sales time for the grain out of storage.

Table 7.3 Rolling a Production Hedge into a Storage Hedge

Cash Position	Futures Position	Basis
May 1		
Crop planted (local cash price $3.60/bu.)	Sell December contract at $3.80/bu.	.20
November 15		
Harvest and put in storage (local cash price $3.00/bu.)	Buy December contract at $3.10/bu.	.10
	+ .70/bu.	change of .10
	Roll into March Sell March contract at $3.50/bu.	.50
March 1		
Sell grain locally at $3.30/bu.	Buy March contract at $3.45/bu.	.15
	+ .05/bu.	change of .35
Net Hedge Selling Price = $3.30 + .70 + .05 = $4.05/bu.		

Likewise, storage hedges can be rolled. This process is quite common with storage hedges. The grain is maintained in storage and continuously hedged with the nearby futures contract, lifted, rolled, lifted, rolled. If the hedge is maintained into the delivery month, the basis offers the best opportunity to be at its most narrow point, as shown in Table 7.4.

Table 7.4 Rolling a Storage Hedge

Cash Position	Futures Position	Basis
December 1		
Put grain in storage (local cash price $3.00/bu.)	Sell March contract at $3.50/bu.	.50
March 1		
Maintain grain in storage (local cash price $3.30/bu.)	Buy March contract at $3.40/bu.	.10
	+ .10/bu.	change of .40
	Roll into May Sell May contract at $3.75/bu.	.45
May 1		
Maintain grain in storage (local cash price $3.50/bu.)	Buy May contract at $3.55/bu.	.05
	+ .20/bu.	change of .40
	Roll into July Sell July contract at $3.80/bu.	.30
July 1		
Sell grain locally $3.60/bu.	Buy July contract at $3.85/bu.	.25
	– .05/bu.	change of .05
Net Hedge Selling Price = $3.60 + .10 + .20 – .05 = $3.85/bu.		

Using Target Prices

Projecting and using target prices offers the potential hedger an excellent management tool. Decisions can be made regarding whether to produce and/or how much to produce. Further, risk-management strategies and pricing alternatives can be evaluated and used if target prices can be formulated. If care is taken in developing basis information, target prices can be one of the best tools producers can use in market analysis.

The first step in using target prices is to have good basis information. Producers are encouraged to keep their own basis information because it allows them to have localized information that is more pertinent to them. Most land-grant universities have basis information for major cash markets, but lack basis information that is localized. Figure 7.2 shows a sample of grain basis published by the University Extension Service at Iowa State University for corn at a major cash market in Iowa. Notice that the average (mean) is shown as well as the range from high to low. Often a median and mode value are helpful as well as a probabilistic

Corn Basis for North Central Iowa
(1989–1990 through 1993–94)
FUTURES CONTRACT MATURITY MONTH

Month & Week	December			March			May			July		
	AVG	MAX	MIN	AVG	MAX	MIN	AVG	MAX	MIN	AVG	MAX	MIN
SEP 4	28	35	20	37	42	29	42	46	35	46	49	39
OCT 1	28	34	21	36	42	30	42	46	36	45	49	41
OCT 2	29	36	18	37	44	27	42	50	33	46	54	38
OCT 3	28	32	24	35	40	32	41	44	37	44	48	39
OCT 4	28	32	25	36	39	31	41	46	35	45	51	37
OCT 5	27	29	25	35	38	31	41	45	34	44	50	37
NOV 1	24	25	22	32	35	28	38	42	32	41	47	35
NOV 2	23	26	21	31	36	28	37	43	31	40	49	31
NOV 3	21	25	19	29	33	26	34	40	28	38	45	27
NOV 4	19	23	17	28	31	25	34	39	28	37	44	27
DEC 1	19	21	17	27	30	26	33	37	28	37	42	28
DEC 2	20	22	17	28	30	25	34	38	30	38	44	30
DEC 3				28	30	24	33	37	29	37	43	29
DEC 4				28	31	24	34	38	30	38	44	31
JAN 1				28	32	26	34	39	29	38	45	28
JAN 2				28	31	24	34	37	31	39	42	35
JAN 3				29	30	26	35	36	34	40	42	35
JAN 4				27	29	26	34	35	33	38	41	34
JAN 5				26	29	25	33	34	30	37	41	31
FEB 1				24	26	22	31	33	27	36	40	28
FEB 2				24	26	21	31	34	27	36	39	28
FEB 3				22	25	21	30	33	28	35	37	31
FEB 4				20	23	19	28	30	26	34	37	29
MAR 1				20	26	17	26	28	24	32	34	27
MAR 2				17	21	14	24	27	21	30	32	24
MAR 3							23	26	20	28	30	25
MAR 4							23	26	21	28	29	24
APR 1							22	27	19	27	29	24
APR 2							21	27	19	26	29	22
APR 3							20	22	18	24	26	21
APR 4							20	26	17	24	28	21
APR 5							18	27	15	23	29	20
MAY 1							19	27	13	23	30	19
MAY 2							20	32	13	23	30	17
MAY 3										23	31	19
MAY 4										23	29	18
JUN 1										23	29	19
JUN 2										23	29	18
JUN 3										23	27	18
JUN 4										23	29	18
JUL 1										23	28	17
JUL 2										23	28	18

Source: Wisner, Robert, *A Composite of Iowa Corn and Soybean Basis Patterns by Price Reporting District 1990–1994*, Iowa State University, University Extension, M-1274, 1994.

Figure 7.2 Weekly average basis corn information for North Central Iowa

value (the probability a certain value will occur), but these values are not normally included in basis tables and are not shown in Figure 7.2. Trying to use target prices without good basis information is essentially a worthless endeavor. Basis is the critical component of target prices; without good basis information the target price cannot be a reliable estimate of net hedge price.

Using the information in Figure 7.2, let's look at how a producer could use target prices in production and market planning. The basis is calculated as

$$\text{FP} - \text{CP} = \text{Basis}$$

for the values in Figure 7.2; therefore, the target price formula

$$\text{TP} = \text{CP} - \text{EEB}$$

will be used. A corn producer is considering planting corn the third week of April. A normal harvest date for the grain is the third week of November. The producer calls her broker and asks for a current quote on December corn futures and is told it is trading at $3.78 per bushel. Producer then looks up in the basis charts what the average basis is for the December contract during the third week of November and finds the average to be $.21 per bushel. This means on average for the period 1989–1994 the December futures price was $.21 cents per bushel higher than the cash price in North Central Iowa. The producer's calculated target price is

$$\begin{aligned} \text{TP} &= \$3.78 - .21 \\ &= \$3.57 \end{aligned}$$

If the producer hedges today at $3.78 and the basis turns out to be $.21 in November, the producer will receive a net hedge selling price of $3.57 per bushel for her corn. The producer may also want to calculate the range that the target price might fall. Figure 7.2 shows the range of basis values for the third week of November to have a high of $.25 and a low of $.19. The target price range would be:

$$\begin{aligned} \text{TP} &= \$3.78 - .19 \\ &= \$3.59 \end{aligned}$$

to

$$\begin{aligned} \text{TP} &= \$3.78 - .25 \\ &= \$3.53 \end{aligned}$$

The range for the producer of the expected net hedged price is $3.53 to $3.59 with an average estimated net hedged price of $3.57.

With this target/expected price information in hand, the producer should be able to make more-informed decisions about the price risk range and what other marketing/pricing alternatives exist and how they compare to the estimated net hedge price (target price). If the producer has an estimate of production costs, then a decision about whether to produce and how much to produce can be made. If the production cost estimate is $3.50 per bushel, then the producer knows that if the worst situation exists for basis in November, then there will be a profit of $.03 cents per bushel ($3.53 – $3.50). The producer may then break down the production costs into cash costs or variable costs. If variable costs are estimated to be $3.00 per bushel, then the producer knows that if the worst basis situation exists, variable costs will be covered as well as fixed costs ($.50 per bushel) and a profit of $.03 per bushel. Similarly, the producer would like to know the probabilities of each target price occurrence. Further basis analysis might indicate that the basis has an 80 percent chance of being $.20 or less. Given that information, the producer would then know that there is an 80 percent chance that she will receive a net hedge price of $3.58 or better.

Target prices can also be used to compare other prices or contract offers. They can serve as a bargaining point or base for forward contracts and can be very useful to producers in formulating marketing goals.

Estimating Production and Operating Costs

For target prices and hedging to be useful to producers, they must have an understanding of their costs. How will a producer know that a target price of $3.57 will cover costs or yield a profit unless they also know their costs of production? As with basis tables, each farmer and rancher must develop a cost accounting system to estimate the cost of production for each commodity produced. The primary purpose of the projected cost-of-production estimate is to determine whether a sufficient profit can be made for the producer to accept the market, production, and other risks associated with each commodity. The process of developing cost-of-production information is not as straightforward as it may seem and deserves a substantial amount of thought. Even the determination of the appropriate cost of production and thus the target price for a farm or ranch that produces only one commodity is not obvious.

Farm and ranch cost-accounting systems will generally take one of two general approaches: (1) They will be an extension of the financial accounting system, or (2) they will be synthesized using a system of input information and formulas to allocate and accumulate costs in various categories. Modern computerized financial accounting packages have opened the door to the development of personalized cost-of-production information directly from the accounting records already being maintained for financial statement and income tax purposes. Individual expense items may be charged to a specific enterprise (commodity) or allocated either individually or according to a predetermined rule to a set of commodities. Though particularly adaptable to cash variable costs, these financial accounting package approaches begin to show difficulties with fixed costs and oftentimes completely collapse with noncash costs and opportunity costs. These approaches are generally an excellent and useful starting point for cost-of-production estimates, yet they usually must be combined with the synthesis approach to be able to consider all costs of production, including economic costs as well as accounting costs.

The synthesis approach concentrates on a routine or process (usually computerized in a spreadsheet or database algorithm) that allocates and categorizes all costs of production so that the producer can make one of several short-run and long-run decisions. Most land-grant institutions develop localized cost-of-production estimates covering the major crops produced in their state using a synthesis approach (often called a budget generator). These cost-and-return estimates (such as the example shown in Figure 7.3) are usually available in printed form. (However, the university may make the budget generator itself available on disk to allow modification for individual farms and ranches.) The printed estimates provide an excellent starting point for producers to use until they have sufficient personalized cost-accounting data for commodities they have produced in the past or for alternative commodities they have not yet tried.

The producer of any commodity must consider not only the variable cash costs of production (such as seed, chemicals, and fertilizer), which must always be covered for production to logically take place, but also unallocated fixed and

CORN FOR GRAIN (figures are per acre)	
	Value or Cost (per acre)
1. Corn receipts	
Corn 120 bushels per acre at $3.00 per bushel	$ 360.00
2. Operating costs	
Seed	15.12
Fertilizer, herbicides, and insecticides	62.89
Crop insurance	9.00
Machinery operating expense	24.68
Labor 2.98 hr. at $5.00/hr.	14.93
Interest on operating capital	5.87
Total operating costs	$ 132.49
3. Income above operating costs	$ 227.51
4. Ownership costs	
Machinery	53.79
Land taxes	12.00
Interest on land investment	88.00
Total ownership costs	$ 153.79
5. Total costs shown	$ 286.28
6. Net returns above costs shown	$ 33.72

Figure 7.3 Sample corn enterprise cost-and-return estimate

variable costs (such as machinery, fuel, and cash land rents) and such noncash costs as unpaid owner/operator and family labor, depreciation on machinery and building, and other opportunity costs. The level at which these costs must be covered with a target price/expected yield in any one year depends upon the type of decision being made.

Producers may wish to estimate whether the target price achievable in the marketplace covers all or a specific portion of the total cost of production. The alternative cost/profit levels include

1. All costs (including opportunity costs) plus a profit margin
2. All costs
3. Variable costs
4. Fixed costs
5. Cash costs
6. Some predetermined percentage of costs

A one-year target price/production decision will usually concentrate on cash variable costs as a minimum for production to take place. Longer-run decisions move upward to levels 1 or 2 because all resources must eventually be replaced.

Selective Hedging

Selective hedging is the process of placing a hedge when price protection is needed and lifting the hedge when the protection is not needed. This is translated as: Producers need to be hedged when prices are falling and not hedged when prices are increasing. This can only be done with accurate forecasts of prices (discussed in Chapter 14).

Consider for example a producer who uses a professional price forecasting service that gives forecasts weekly on price outlooks for the next few weeks. The producer has found the service to be accurate more often than not. The service puts out the following recommendation: "Keep inventories and do not sell during the next two weeks, as prices are likely to firm up as the news of a lighter than average test weight crop hits the marketplace" (translation—prices will increase during the next two weeks).

A selective hedger might consider remaining unhedged and reconsider hedging as the service keeps information current on possible price decreases. The upside of selective hedging is, of course, that you are not hedged when prices are moving in favor of your cash position. The downside is that forecasts are not perfect and accurate 100 percent of the time. You may find yourself unhedged when you need to be.

Selective hedging can also be done using basis. A hedge is placed (a futures contract is sold) when the basis is widest and lifted when it is the narrowest and/or when the cash commodity is sold. Looking at Figure 7.2 once more, the corn producer may want to sell the corn not when harvested during the third week of November, but during the first or second week of December when the basis is smaller. This might involve a few weeks of storage, but it could also lead to a later planting date next year or a shift to a later maturing variety.

Merchandisers

Most grain and oilseed merchandisers use futures to hedge their purchases and sales. Merchandisers are in the business of buying and selling grains and oilseeds, and consequently they usually have a cash position that has price risk exposure. Merchants may buy and then later sell, but they also often sell first and then buy to cover the forward sale. They can therefore be both long and short hedgers under different circumstances.

Short Merchandising Hedges

The more traditional situation involves a cash purchase and then a later sale. During the time the merchant is long the cash, the risk is that prices will decrease and the commodity will be sold at a loss. This can be protected by a short hedge, as illustrated in Table 7.5. The merchandiser wants the basis to narrow or improve to capture additional revenue from the hedge. If properly hedged, the merchandiser does not care whether prices increase or decrease because the net hedge selling price will remain the same. In the example in Table 7.5, the merchandiser had a gross profit of $.02 per bushel because there was a favorable basis movement regardless of price direction.

Table 7.5 Short Merchandising Hedge

Cash Position	Futures Position	Basis
November 15		
Buy corn at $3.00/bu.	Sell December contract at $3.21/bu.	.21
	Price Decrease	
November 25		
Sell corn at	Buy December contract at	
$2.90/bu.	$3.09/bu.	.19
– .10/bu.	+ .12/bu.	change of .02
	Net Hedge Price = $2.90 + .12 = $3.02/bu.	
	Price Increase	
November 25		
Sell corn at	Buy December contract at	
$3.10/bu.	$3.29/bu.	.19
– .10/bu.	– .08/bu.	change of .02
	Net Hedge Price = $3.10 – .08 = $3.02/bu.	

Long Merchandising Hedges

Oftentimes it is possible for a merchant to forward sell the commodity for delivery in the future before a cash purchase has been made. The merchant is short the cash and needs to be protected with a long futures position. The merchant has the risk with a short cash position that prices will increase and the grain will have to be purchased at a higher price than it was sold. An example is presented in Table 7.6 with a basis improvement. Notice that the improvement in basis for the long hedge is a widening from the initial position. Again, the properly hedged merchant does not care what direction prices go, only what basis does.

Table 7.6 Long Merchandising Hedge

Cash Position	Futures Position	Basis
November 20		
Sell corn for delivery in two weeks at $3.20/bu.	Buy December contract at $3.40/bu.	.20
	Price Decrease	
November 30		
Buy corn at $3.10/bu.	Sell December contract at $3.32/bu.	.22
+ .10/bu.	− .08/bu.	change of .02
	Net Hedge Price = $3.10 − (−.08) = $3.18/bu.	
	Price Increase	
November 30		
Buy corn at $3.30/bu.	Sell December contract at $3.52/bu.	.22
− .10/bu.	+ .12/bu.	change of .02
	Net Hedge Price = $3.30 − (.12) = $3.18/bu.	

Basis Aspects

Merchandisers are very aware of the importance of the basis in whether they have profits or losses. Many have developed into basis traders, and much of the grain traded is traded using basis terminology. A merchant may offer to buy grain "3 cents off," meaning that he is willing to pay three cents under the nearby futures price, or perhaps "2 cents on," meaning two cents over the nearby futures price. Merchandisers are constantly making offers to buy or sell cash grain expressed in prices relative to the futures price. This accomplishes two purposes. First it is shorthand verbiage and necessary in the grain and oilseed trading environment where most trades are by phone or fax. Secondly it allows the trader to mentally fix what basis values are necessary for a profitable trade. They don't have to worry about price levels, price direction, or the actual cash price. They merely concentrate on two values: the buying basis and the selling basis (or the beginning and ending basis values).

If a basis-trading merchant buys corn $.02 under the nearby and then sells it for $.01 under the nearby, he has made a gross return of $.01 per bushel (shown in Table 7.7). A solid understanding of the grain and oilseed market, futures markets, and especially basis is necessary for a merchant to successfully basis trade. Many grain companies require several years of apprenticeship and training before they allow their merchandisers to handle large positions.

Table 7.7 Basis Trading Example

Cash Position	Futures Position	Basis
November 11		
Buys grain at "2 under the nearby" thus a cash price of $3.78/bu.	Sells December contract at $3.80/bu.	.02
November 12		
Sells grain at "1 under the nearby" thus a cash price of $3.75/bu.	Buys December contract at $3.76/bu.	.01
− .03/bu.	+ .04/bu.	change of .01
Net Hedge Selling Price = $3.75 + .04 = $3.79/bu.		

To-Bid Price

A class of merchandisers known as elevators buy grain and oilseeds from producers. They quote price on a regular schedule (usually at least daily). The quoted cash price must be determined by some rational means (guessing is not a consistently successful process!). The normal rational means used by many elevators is called the **to-bid price**.

A to-bid price is very similar to a target price. It involves taking the nearby futures price and subtracting (and sometimes adding) variables. The variables usually subtracted are transportation charges, carrying charges, and profit. For some locations an additional differential is added such as the "Gulf-Tulsa" differential to reflect regional supply-and-demand situations.

A grain elevator in central Iowa would take the nearby futures quote today, subtract the transportation charge from central Iowa to the par delivery point for the corn futures (Chicago), subtract the carrying charges to reflect the difference between the date today and the delivery time for the futures contract, and subtract a profit margin (shown in Table 7.8). This is a rational cash quote because if the grain elevator bought the grain for $3.05 per bushel and hedged it at $3.40 per bushel, then the elevator could store the grain until December (costing $.15), deliver against the futures contract in Chicago (costing $.18 in transportation and elevation charges), and make $.02 profit. Although storage and transportation costs cannot be forecasted with perfect accuracy, their costs have much less variability than do grain prices. Obviously the amount of profit an elevator can capture depends upon local competition.

Table 7.8 Example of "To-Bid Pricing"

TBP = CFP – T – CC – P

where

TBP	=	To-Bid Price
CFP	=	Current Nearby Futures Price
T	=	Transportation Charges
CC	=	Carrying Charges
P	=	Profit

Example:

CFP	=	$3.40/bu.
T	=	– .15/bu.
CC	=	– .18/bu.
P	=	– .02/bu.
TBP	=	$3.05/bu.

Not all elevators in an area will to-bid price. Most will price based on what their competition is doing. Usually a larger elevator or a price leader elevator will to-bid price and other elevators in the area will base their prices upon the price leader elevator. Small adjustments will be made in the price by each elevator relative to the price leader. It must be noted, however, that only the elevator that hedges the cash purchases will be able to capture the actual to-bid price. If an elevator has a cash quote of $3.00 per bushel based upon a price leader's to-bid price, the elevator must find a buyer for the grain at $3.00 per bushel or above. If no buyer is found and the elevator is unhedged, then the to-bid price cannot be received. The hedged elevator, however, if they do not find a buyer, can always deliver against the futures and receive the to-bid price.

The elevator that does not hedge and uses to-bid pricing can only be buying and selling back-to-back and probably won't exist long in today's highly competitive grain business. If they don't buy and sell back-to-back or sell and buy back-to-back, then they are buying and holding and speculating that the price will increase. They may make a profit by speculating, but the long-term outlook for such an elevator is not promising.

Spread Hedging

A spread hedge is simply a spread (Chapter 7) from a mechanical standpoint and a hedge from the standpoint of trying to protect a relative value. Normal hedges protect against absolute price changes, and spread hedges protect relative values between two or more markets. Grain elevators can watch the spread between months and roll their storage hedge into a more distant month that earns better storage income.

A widely used spread hedge is a process used in the soybean market called crush and reverse crush.

Crush

Soybeans are processed initially into two products: soybean meal and soybean oil. The original process was a crushing mill that extracted the oil from the meal. Modern plants use solvents to extract oil from the meal, and then the solvent is boiled off to leave pure oil. The system is closed, and the solvent is reused. Although the process is no longer a physical crush of the soybean, it is still referred to as crush. A typical 60-pound bushel of soybeans will yield about 11 pounds of oil, 48 pounds of meal, and 1 pound of waste. This is referred to as the crush yield. The crush yield varies with varieties, growing conditions, and crush processes. The crush yield is reported on a regular basis by the United States Department of Agriculture.

A soybean processing plant buys soybeans and processes them into meal and oil. They have the price risk that soybean prices will increase and the prices of soybean oil and meal will decrease and thus the crush margin will be reduced until the equilibrium level is reached. Their risk is really one of relative price movements (spreads) between soybeans and soybean meal and oil. They can use a spread hedge to protect the crush margin.

To get started, convert the oil and meal into equivalent bushel values (the soybean oil and meal futures contracts are standardized using tons). Table 7.9 illustrates the process. If the value of the meal and oil is greater than the cost of the soybeans plus the cost of crush, then a spread hedge called putting on crush is initiated (shown in Table 7.10). The processor will capture the crush margin using this process. The action of buying the soybeans will have a tendency to push soybean prices up, and the action of selling the meal and oil will tend to drive prices down. The price of soybeans plus what it costs to process them into meal and oil should equal the value of the meal and oil. This is called the normal relationship. It is when this normal relationship does not exist or does not remain constant that profit opportunities exist.

Table 7.9 Soybean Crush Spread Hedge Equivalent Bushel Calculations

Crush Yield: 60-pound bushels of beans yield 48 pounds of meal, 11 pounds of oil, and 1 pound of waste[1]

November soybean futures trading at $7.30 per bushel
November meal futures trading at $.10 per pound
November oil futures trading at $.25 per pound

Value of meal and oil = $7.55
Meal (48 pounds × .10) = $4.80
Oil (11 pounds × .25) = $2.75

Cost of crush = $.20/bu.

[1] Crush yield varies with the quality of soybeans. Many processors yield only 44 pounds of 49 percent protein meal, but the 48-11-1 ratio is a standard that is used here to illustrate the crush spread hedge.

Table 7.10 Example of Crush Spread Hedge

From Table 7.9:

Value of November meal and oil = $7.55/bu.
Cost of crush = $ .20/bu.
November soybeans = $7.30/bu.

Crush Rule: If VMO (Value of meal and oil) is greater than SP (soybean price) + CC (Crush Costs), then Buy Soybean Futures and Sell Soybean Meal and Oil Futures.

Spread Hedge: Putting on Crush

Leg A: Buy Soybean Futures at $7.30/bu.
Leg B: Sell Soybean Meal and Oil Futures at $7.55/bu.

Markets should go back to normal, which is the cost of crush ($.20). When this happens, as:

November soybean futures at $7.32/bu.
November soybean meal and oil at $7.52/bu.

then the spread hedge is lifted:

Leg A: Sell Soybean Futures at $7.32/bu.
Leg B: Buy Soybean Meal and Oil Futures at $7.52/bu.

Net Leg A: + $.02/bu.
Net Leg B: + $.03/bu.

Net hedge addition to crush margin = $.05/bu.

Soybean processors can take advantage of abnormal relationships and add to the crush margin or they may spread hedge when the relationship is normal to protect themselves so that they can be assured of covering the cost of crush.

Reverse Crush

When the cost of the soybeans plus crush costs are greater than the value of the meal and oil, a reverse crush can be placed. A reverse crush involves selling soybean futures and buying meal and oil futures. This action, if done by many traders, will drive soybean prices down and meal and oil prices up to the equilibrium level at which the market comes back to normal. When it does, the spread hedge is offset and the change in the spread is captured for the hedger. A reverse crush spread hedge is illustrated in Table 7.11.

Table 7.11 Example of Spread Hedge Reverse Crush

Value of March meal and oil = $7.20/bu.
March Soybeans = $7.10/bu.
Cost of Crush = $ .20/bu.

Reverse Crush Rule: If VMO SP + CC, then put on reverse crush, as: Sell Soybean Futures and Buy Meal and Oil Futures

Spread Hedge: Putting on Reverse Crush

Leg A: Sell March Soybeans at $7.10/bu.
Leg B: Buy March Meal and Oil at $7.20/bu.

Markets should go back to normal, which is the cost of crush ($.20). When this happens, as:

March soybean futures at $7.08/bu.
March meal and oil futures at $7.28/bu.

then the spread hedge is lifted:

Leg A: Buy March Soybeans at $7.08/bu.
Leg B: Sell March Meal and Oil at $7.28/bu.

Net Leg A: + $.02/bu.
Net Leg B: + $.08/bu.

Net hedge addition to crush margin = $.10/bu.

Both crush and reverse crush spread hedges operate on the premise that the market is temporarily abnormal and will move towards normality. But even if it doesn't, the crush hedger will still be protected and will capture the difference because he is long the soybeans. Delivery can be accepted, the beans crushed, and then the meal and oil delivered against the short futures position. However, the reverse crush position is not as well protected. The hedger is short soybeans and long meal and oil, so he would deliver and accept delivery of meal and oil—not what he is in business to do. Putting on crush hedges assures the processor of capturing the crush margin difference regardless of whether or not the market goes back to normal because the processor can go through delivery. Reverse crush hedges will capture the crush margin difference for the hedger only when the market goes back to normal, otherwise going through delivery will not capture the margin difference. Both the crush and reverse crush spreads are illustrated and compared in Table 7.12.

Table 7.12 Comparison of Crush and Reverse Crush

Crush:

Leg A:	Buy November Soybeans at $7.30/bu.
Leg B:	Sell November Soybean, Meal, and Oil at $7.55/bu.

If the markets don't go back to normal, ($.20), then in November the processor can accept delivery of the soybeans and pay $7.30/bu., process them into meal and oil at a cost of $.20, and then deliver against the futures and receive $7.55.

Net Crush Margin = $.25/bu.

Reverse Crush:

Leg A:	Sell March Soybeans at $7.10/bu.
Leg B:	Buy March Meal and Oil at $7.20/bu.

If the markets don't go back to normal, then the delivery process will not assure the processor of an increase in the crush margin. If the processor went through delivery, he would receive meal and oil and deliver beans and would not, through this action, receive an increase in the crush margin.

Ex-Pit Transactions

Ex-pit transactions are also known as exchange of futures (EOF). EOF is a process of allowing the offsetting transactions of a hedge to be done outside the normal open outcry auction of the trading pits and simply involves an exchange of the futures positions of the two trading partners. For an ex-pit transaction to be allowed, the two trading parties must be hedged opposite and there must be a corresponding cash transaction.

Ex-pit transactions are usually made by grain and oilseed merchandisers trading basis. They almost always entail large volumes of grain and oilseed contracts. Consider as an example a grain exporter and a regional grain merchandiser. The grain merchandiser has grain purchased from producers at an average price of $3.00 per bushel and hedged with a March corn futures at $3.30 per bushel. The exporter has found an overseas buyer and enters into a forward sale of corn at a price of $3.35 per bushel and has it hedged at $3.45 with the March corn futures. The exporter needs grain to fulfill the contract, and the grain merchandiser wants to sell grain. They agree to a sale/purchase at $.15 off the March futures, ex-pit. The traders' brokers inform the exchange, and the clearing corporation takes the closing price for the day (although it doesn't matter which price they choose) for the March futures. The exchange of the futures is recorded, and the traders do not have to issue futures buy and sell orders. The transactions are shown in Table 7.13.

Table 7.13 Ex-Pit Transactions Example

EXPORTER		
Cash Position	**Futures Position**	**Basis**
February 1 Forward sells corn for delivery in two weeks at $3.35/bu.	Buys March at $3.45/bu.	.10
Agree to exchange and trade at 15 off March		
Exchanges at Closing Price, March Futures		
Buys at $3.33/bu. + .02	$3.48/bu. + .03	.15 change of .05
Net Hedge Buying Price = $3.33 – (.03) = $3.30/bu.		

REGIONAL GRAIN ELEVATOR		
Cash Position	**Futures Position**	**Basis**
Has grain in storage purchased at $3.00/bu.	Sell March at $3.30/bu.	.30
Sells at $3.33/bu.	Exchanges at $3.48/bu. – .18	.15 change of .15
Net Hedge Selling Price = $3.33 – .18 = $3.15/bu.		

Both the exporter and the grain elevator had positive gross margins. They agreed upon a basis ($.15) that caused both to have an improvement ($.05 for the exporter and $.15 for the elevator). The trade was beneficial to both parties. By exchanging futures, they avoided having to offset their hedges through normal trading and also fixed a single price at which they exchanged their contracts.

Call Contracts

Call contracts consist of buyer's calls, seller's calls, and doubles. A call contract is an agreement to buy or sell the cash commodity at a specific basis. A call contract must have a limited duration and list a specific futures month as the base contract. The two trading parties must agree on a basis.

Seller's Call

A producer and an elevator enter into a seller's call by agreeing to allow the producer to select a pricing date at her discretion (within limits) and at specified price differential relative to the futures. For example, assume that the elevator agrees to let the producer have any day's price for the March corn contract less ten cents up until March 1. The elevator gets title to the producer's grain now and the producer gets to price it anytime between now and March 1. The producer will receive the March corn futures price less ten cents. The elevator has the risk that corn prices will increase between now and March 1, and the producer has the risk that the price will decrease.

The producer will call the elevator's broker the day she wants to price the grain and tell the broker to sell a corn futures contract for the elevator (thus the name seller's call). If the elevator sold the grain when the contract was entered, then he long hedged the grain, and the seller's call offsets the hedge. If the elevator did not sell the grain and still has it in inventory and unhedged, then the seller's call puts the elevator into a short hedge to protect against price decreases. Table 7.14 shows both these situations. The elevator will be trying to sell the grain as quickly as possible; thus, most seller's calls are offsetting a previous long hedge for the elevator unless they are called fairly quickly.

Table 7.14 Example of Seller's Call for an Elevator

FORWARD SALE		
Cash Position	**Futures Position**	**Basis**
November 15		
Sells grain at $3.90/bu.	Buys March at $4.00/bu.	.10
February 15		
Buys grain at $4.40/bu.	Producer calls elevator's broker and has a March contract sold at $4.50/bu.	.10
– .50/bu.	+ .50/bu.	No change
Net Hedge Buying Price = $4.40 – .50 = $3.90		

NO FORWARD SALE		
Cash Position	**Futures Position**	**Basis**
November 15		
Contract entered into		
February 15		
Buys grain at $3.65/bu.	Producer calls elevator's broker and has a March sold at $3.75/bu.	.10

The producer could likewise hedge against price declines and could possibly have a basis improvement, as illustrated in Table 7.15. The producer gained the basis change of $.05 by hedging the seller's call; however, because she was hedged, she gave up the $.35 increase in the cash market.

Table 7.15 Example of Seller's Call for a Producer

Cash Position	Futures Position	Basis
November 15		
Local Cash Price at $3.30/bu.	Sell March Futures at $3.45/bu.	.15
February 15		
Calls elevator's broker and gets the March futures price of $3.75 less 10 cents/bushel.		
Sells $3.65/bu.	Buys March at $3.75/bu.	.10
	– .30/bu.	change of .05
Net Hedge Selling Price = $3.65 – .30 = $3.35/bu.		

Buyer's Call

Conceptually a buyer's call and a seller's call are the same except that opposite futures positions are taken. For example, a processor needs grain and arranges with an elevator to transfer the grain to him and enters into a buyer's call. They agree on a basis of ten cents under the March futures, and the processor has until March 1 to price the grain. The elevator has the grain purchased and hedged at a basis of $.15. When the processor is ready to price the grain, he calls the elevator's broker and has the broker buy a March futures contract for the elevator, which offsets the elevator's short hedge. The elevator had a gross margin of $.05, as illustrated in Table 7.16. Regardless of what happens to prices, the elevator will profit by the five-cent change in the basis because both bases are fixed.

Table 7.16 Example of Buyer's Call for An Elevator

Cash Position	Futures Position	Basis
November 30		
Grain in storage purchased at $3.90/bu.	Sell March at $4.05/bu.	.15
Two weeks later the manufacturer calls and has the elevator's broker buy a March contract at $3.70/bu.		
Sells grain at $3.70/bu.	Buy March at $3.80/bu.	.10
– .20/bu.	+ .25/bu.	change of .05
Net Hedge Selling Price = $3.70 + .25 = $3.95/bu.		

The processor could hedge to protect against increasing prices by being a long hedger and establishing a beginning hedge basis smaller than ten cents under such as five cents under, thus gaining a five-cent basis gain.

Double Calls

An elevator could simultaneously enter into both a buyer's call and a seller's call. This fixes both bases for the elevator and if done properly, will allow the elevator to guarantee a basis improvement. The simple rule is that the elevator wants the seller's call basis to be larger than the buyer's call basis. If the elevator entered a seller's call for a ten-cent-under-the-March basis, then she would want to enter a buyer's call with a basis that is less than $.10, such as $.05. The elevator then does nothing. If the seller's call is enacted first, the elevator will be hedged short at a $.10 basis, and then when the buyer's call is activated, it will offset the short hedge at a basis of $.05, thus creating an improvement of $.05. Similarly, if the buyer's call was enacted first, the elevator would be hedged long at a basis of $.05 and would then offset with the seller's call at $.10 and an improvement in the basis of $.05. *It doesn't matter which call is enacted first.* All that matters is that the proper basis difference was established in the initial opening position. The double call example is illustrated in Table 7.17.

Table 7.17 Example of Double Call

BUYER'S CALL ENACTED FIRST		
Cash Position	**Futures Position**	**Basis**
Sell at $3.55/bu.	Buy March at $3.60/bu.	.05
Buy at $3.60/bu.	Sell March at $3.70/bu.	.10
− .05/bu.	+ .10/bu.	change of .05
Net Hedge Buying Price = $3.60 − .10 = $3.50		
SELLER'S CALL ENACTED FIRST		
Cash Position	**Futures Position**	**Basis**
Buy at $3.65/bu.	Sell March at $3.75/bu.	.10
Sell at $3.50/bu.	Buy March at $3.55/bu.	.05
− .15/bu.	+ .20/bu.	change of .05
Net Hedge Selling Price = $3.50 + .20 = $3.70/bu.		

Basis Contracts

Basis contracts are also known as price later contracts, and they are also similar to call contracts. The only real difference between basis contracts and call contracts is that the elevator or contracting party handles the actual buying and selling of the futures contracts rather than the other contracting party as with call contracts.

A basis price later contract is common between producers and elevators in certain areas of the United States. The mechanics are the same as call contracts. A producer and elevator agree upon a basis, futures contract month, and expiration date. The producer is allowed to speculate that grain prices will increase without actually holding the grain, and the elevator receives title so that it can merchandise the grain. Recognize that the producer is speculating and thus is not managing price rise with a basis contract unless he hedges. The elevator must hedge the contract to avoid the risk of price increases if the elevator sells the grain before the producer requests a price. If the elevator does not sell the grain before the contract is priced, then the elevator will short hedge once the buying price has been established.

Several variations exist for basis contracts. Some require the producer to pay for storage and/or a one-time elevation charge. Each must be evaluated based upon the specific contract terms and the levels of risks for the producer. During the early 1990s, basis contracts called "Hedge to Arrive" emerged. They are mechanically no different than a single basis contract.

Basis Aspects

Grains and oilseeds are storable commodities. The difference between a cash price at one location and the futures price should be the cost of carry (known as COC). The COC is composed of the cost of storage and the opportunity cost of money. The cost of storage is made up of the labor, cost of the physical facility and insurance, and all other costs needed to hold the grain or oilseed in a storage location. The opportunity cost of money is the cost associated with keeping the grain rather than converting it to cash and investing the cash.

If the location of the grain is not a delivery point for the futures contract (highly likely), then a freight charge is also part of the difference between the cash price and the futures price. The COC and freight are sometimes referred to as CFIs—Costs, Freight, and Insurances. The Costs are all physical costs associated with storing the grain and also the elevation charges, which are the cost of moving grain in and out of storage. Freight is the cost to move the physical grain to a different location. Insurances refers to the various costs of insurance to hold and ship grain. Typically COC is the term used for domestic storage of grains, whereas CFI is the term preferred for international shipments of grain.

Because the cost of carry, freight, and insurance is fairly constant, the basis pattern for grain and oilseed is well established. The COC erodes over time, thus the basis value for grain decreases over time. Consider for example corn in storage in March of 1998 with a spot price of $3.00 per bushel. What should the value of the December corn futures contract be for 1998? It should reflect the cost to store the grain from March to December plus the freight difference between where the grain is stored and the closest par futures delivery point. The freight cost will be a constant at say $.05 per bushel while the COC erodes each month by the costs per month, say $.06 per month per bushel. The storage from March to December is nine months at $.06 equals $.54 per bushel. In April, the total storage cost will have eroded by the cost of one month's storage to $.48 per bushel. The December corn futures should be trading $.54 per bushel above the local cash price in March. As time passes, the cash price will move closer to the futures price.

PROBLEMS

1. Soy, Inc. (a soybean processor) observes that November soybean futures are at $8.50 per bushel, November meal futures are at $.16 per pound, and November oil futures at $.12 per pound. Crush costs are running $.35 per bushel. What should Soy, Inc. do? What is their potential gain?
2. Congress Grain Company has purchased grain and hedged it with a basis of $.06 per bushel. For Congress to have a positive gross revenue, should they sell the grain at a basis of $.03 or $.09 per bushel? Explain.
3. Murry is a grain merchandiser. He has negotiated a seller's call with a client at $.10 per bushel and a buyer's call with another client for $.07 per bushel. Was this a favorable deal to Murry? Explain.
4. Tatianna puts 10,000 bushels of wheat into storage. The current cash price is $4.00 per bushel. She hedges the stored wheat with two December wheat futures at a price of $4.25 per bushel. On November 25 Tatianna rolls the hedge. She offsets her December contracts at $4.00 per bushel and rehedges with two March contracts at $4.15 per bushel. The cash price at the time she rehedged was at $3.98 per bushel. On February 15 Tatianna sells the grain out of storage in the cash market for $4.20 per bushel and lifts the hedge at a price of $4.22 per bushel. What is Tatianna's net hedge price? Show and explain.

Hedging Strategies: The Livestock and Meat Complex

"Never invest in anything that eats or needs repairing."

—Billy Rose

The livestock futures markets complex consists of feeder cattle, live (market-ready fed) cattle, hogs, pork bellies, cheddar cheese, nonfat dry milk, fluid milk, butter, boneless beef, and shrimp just to name a few. Over the years various other livestock products have also had futures contracts, including frozen beef, broilers, wool, and eggs. Lack of trading activity caused most of them to be dropped. The current cattle, hog, and pork belly contracts have existed since the 1960s and are viable contracts. Futures contracts on two dairy products, cheddar cheese and nonfat dry milk, were introduced by the New York Coffee, Sugar, and Cocoa Exchange in June 1993. White shrimp contracts started trading on the Minneapolis Grain Exchange in 1993. In 1994 the Black Tiger shrimp contract was listed. The Chicago Mercantile Exchange added boneless beef, boneless beef trimmings, and ground beef in 1997.

Producers

Livestock producers are not as easily defined as grain producers because livestock do not have growing periods, with more definite seasonal beginning and ending dates, that are as inflexible as grain. For example, cattle producers include cow-calf operations that perform the breeding, gestation, and delivery of calves. Although traditionally calves are born in the spring, this process actually can and does continue all year. Cattle production also includes the process of providing the time and an environment for calves to put on additional weight. The livestock production process can be broken up several times by different firms providing different aspects of the growing process. In fact, it is very rare for one firm to perform all the functions from breeding to slaughter.

For cattle, two futures contracts can be used for hedging: the feeder cattle contracts (650-pound animals) and the live (fed) cattle contract (1,000-pound animals). The Chicago Mercantile Exchange has listed boneless beef, boneless beef trimmings, and ground beef in 1997 to expand into the consumer beef business. Hogs have only one live market hog contract (220-pound animals) and one final meat product, pork bellies (uncured bacon). Each of the livestock contracts calls for approximately a semitruckload unit (live cattle contracts are denominated in 40,000-pound units; feeder cattle, 50,000 pounds; and lean hogs, 40,000 pounds).

Production Hedges

The price risk in livestock production is the same as for grain and oilseed producers—the risk that prices will decrease during the production process. Consequently, livestock production hedges are short hedges.

Live Cattle

Live cattle are a product of a cattle finishing operation that feeds a concentrated protein ration. Typically this is done in large confinement operations called feed yards or feedlots. The feeding period varies from short periods (less than 60 days) to very long periods (greater than 180 days). Typically, most cattle feeding periods are from 150 to 165 days. Feeding periods vary in length due to the age, condition, in-weight and breed of the cattle, feeding conditions (ration composition and time of feed), weather, and health conditions. The live cattle futures contract calls for all choice grade steers (castrated males) that average 1,000 pounds per animal. Consequently, a feeder of heifers or of good grade cattle (such as finished cattle of dairy breeds) actually uses a form of hedging called a cross hedge when using the live cattle futures contract. This process is discussed in more detail in Chapter 11.

Live cattle hedges are either anticipatory hedges or purchased hedges. An anticipatory hedge is one that is placed by someone that has the cattle in a growing program and anticipates that they will put them in a finishing operation. A purchased hedge is placed by someone who buys cattle that are ready to be put into a finishing operation.

An anticipatory hedge typically involves someone who has feeder cattle either on range or permanent grass or on planted pasture (generally fall-planted wheat, rye, or other small grain) and who plans to move them into a finishing operation and "feed them out." The risk is that while the cattle are growing and being finished in the feed yard, the price of live (fed) cattle will decrease, sometimes far enough to eliminate all profit margins. Table 8.1 shows an anticipatory live cattle hedge. For illustrative purposes, a basis deterioration is shown. A basis improvement is possible if the basis values were the opposite of those shown.

Table 8.1 Anticipatory Live Cattle Hedge

Cash Position	Futures Position	Basis
January 1		
Have cattle in feedyard (current fed cattle price $61.00/cwt.)	Sell June live cattle contract at $62.50/cwt.	$1.50
May 1		
Cattle finished and sold at $58.00/cwt.	Buy June live cattle contract at $60.00/cwt.	$2.00
	+$2.50/cwt.	change of $.50
Net Hedge Price = $58.00 + $2.50 = $60.50/cwt.		

A purchased hedge is very similar to the anticipatory hedge. The only real difference is that the cattle are purchased and immediately put on feed. The only trick to this type of hedging is calculating basis from the correct values. In Table 8.2 the correct buying basis (beginning basis) is calculated from the live cattle price, not the price paid for the feeder cattle. Calculating it using the feeder cattle purchase price leads to incorrect net hedged selling prices and misleading basis analysis. The example in Table 8.2 for the purchased hedge shows a basis improvement for the feeder.

Table 8.2 Purchasing Live Cattle Hedge

Cash Position	Futures Position	Basis
January 1		
Buy feeder cattle at $70.00/cwt., place in feedyard (fed cattle price $61.00/cwt.)	Sell June live cattle contract at $62.50/cwt.	$1.50
May 1		
Cattle finished and sold at $58.00/cwt.	Buy June live cattle contract at $59.00/cwt.	$1.00
	+$3.50/cwt.	change of $.50
Net Hedge Price = $58.00 + $3.50 = $61.50/cwt.		

In addition to the owner of the cattle hedging, many feedlots provide hedging services. The feedlot itself will hedge the live cattle for the producer. Some of these services merely provide hedging services, whereas others provide an integrated and detailed marketing plan for feeding cattle that involves estimating costs, basis, and determining what if any profit potential exists. These marketing plans represent a very sophisticated approach to not only hedging for price risk protection but also for determining if feeding cattle is profitable and should be undertaken.

Hogs

Hog producers either produce hogs from farrow (birth) to finish or buy feeder pigs at approximately 50–75 pounds and feed them to market weights usually in the range of 210 to 230 pounds each. Hog feeding periods vary from three to four months typically. The feeding period varies depending upon age, condition and the breed of the hogs, health conditions (very important for feeder pigs), and feeding conditions (ration, timing, weather, and shelter). Hog contracts allow both barrows (castrated males) and gilts (nonpregnant females) that average 220 pounds. Because there is no feeder pig futures contract, only the final product—live market hogs—can be hedged. Hog production hedges fall into the same category of cattle hedges—anticipatory and purchased. Because hog production hedges are mechanically the same as cattle hedges, only the anticipatory hedge is illustrated with an example (Table 8.3).

Table 8.3 Anticipatory Hog Hedge

Cash Position	Futures Position	Basis
January 1		
Have pigs on feed (hog price at $45/cwt.)	Sell April hog contract at $48/cwt.	$3
March 15		
Hogs finished and sold at $40/cwt.	Buy April hog contract at $42/cwt.	$2
	+$6/cwt.	change of $1
Net Hedge Price = $40 + $6 = $46/cwt.		

Feeder Cattle

Production hedges for feeder cattle (650-pound choice steers) are typically used by two types of producers: cow-calf producers who carry the calves to feeder cattle weights, and feeder cattle producers. Cow-calf production involves breeding, gestation, and growth of the animals from calves (at birth most calves weigh less than 75 pounds) to feeder cattle weights (approximately 650 pounds). Feeder cattle producers are not involved in the breeding, gestation, and preweaning portion of the production process. Rather, they purchase animals from as light as 250 pounds to approximately 500 pounds (a typical weaning weight). Cattle feeders may bypass the feeder cattle producers to purchase lightweight calves (250–550 pounds) or they may opt for shorter feeding periods and purchase feeder cattle.

Tables 8.4 and 8.5 show examples of the cow-calf producer and the feeder cattle producer hedges. There is no major difference between the two hedges. The cow-calf producer hedge is very much an anticipatory hedge, and the feeder cattle producer hedge is a purchasing hedge. A basis improvement is shown for the cow-calf producer hedge, and a basis deterioration is shown for the feeder cattle producer hedge.

Table 8.4 Cow-Calf Producer Feeder Cattle Hedge

Cash Position	Futures Position	Basis
May 1		
Calving completed and calves weigh about 200 lbs. (feeder cattle price at $70/cwt.)	Sell April feeder cattle futures at $72/cwt.	$2
March 15 of next year		
Calves now are feeders (weighing 650 lbs.). Sell at $60/cwt.	Buy April feeder cattle futures at $61/cwt.	$1
	+$11/cwt.	change of $1
Net Hedge Price = $60 + $11 = $71/cwt.		

Table 8.5 Feeder Cattle Producer Hedge

Cash Position	Futures Position	Basis
June 1		
Buys 250-pound animals at $90/cwt. (feeder cattle price at $68/cwt.)	Sell April feeder cattle futures at $70/cwt.	$2
March 15		
Calves now are feeders (weighing 650 lbs.). Sell at $60/cwt.	Buy April feeder cattle futures at $63/cwt.	$3
	+$7/cwt.	change of $1
Net Hedge Price = $60 + $7 = $67/cwt.		

There are several producers who do not fall into either of these categories. Many cow-calf operators sell their animals at light weights or as stockers (350+ pounds). Several firms and individuals buy lightweight animals, put them on pasture to increase body frame size but do not feed grain for weight gain, and sell them as stockers or at about 500 pounds. These producers are not producing feeder cattle (defined by the contract to average 650 pounds and fall within a 10-pound range) and therefore cannot exactly match the feeder cattle futures contract.

Furthermore, the feeder cattle contract calls for steers (male castrated animals) with no allowance for heifers (female animals that have not had a calf). Mother nature produces approximately equal numbers of sexes in both cattle and hogs, and this confounds the hedging process. These problems are not insurmountable. They can be handled with a cross hedge (Chapter 11).

Beef Products

The Chicago Mercantile Exchange's new beef products futures offer meat packers a new way to hedge their output. Prior to the beef products contracts, packers had to speculate on final output prices and/or cross hedge with live cattle futures, which proved to be a very poor cross hedge and consequently not widely used. The new contracts cover lean beef, lean beef trimmings, and ground beef.

Two major types of hedging opportunities exist for beef packers. First, the quantities of beef that exist in a carcass can be determined so that a packer could match the output of product relative to the cattle input. Or secondly, a packer could match total output of each beef product to the overall processing plant output. The first example is very similar to a soybean processor crush problem because it is really a spread hedge; the second example is a simple production hedge.

A market-fed steer weighing 1,000 pounds live, will dress out at approximately 60 percent (i.e., produce a carcass of about 600 pounds—the remainder of

400 pounds is called offal and includes the hide, body fluids, and internal organs and parts). The carcass is then broken down into retail cuts of meat and will yield approximately 70 percent, or about 420 pounds of retail beef products, lean beef trimmings, and ground beef (the remainder of 180 pounds is moisture loss, bones, fat, and other tissues). The 580 pounds of non-lean beef products has value—some at retail such as tongue, liver, and tripe. The offal and bones are used to make other nonfood products. The hide is sold into the leather tanning market.

A beef packer can now put a value on each of the items and determine if it is profitable to process beef, such as

1,000 pound live animal at $.79/lb.	=	$790
Cost of processing ($35/head)	=	$ 35
Total cost	=	$825
Value of Products		
400 pounds of hide/offal at $.30/lb.	=	$120
180 pounds other at $.20/lb.	=	$ 36
430 pounds retail meat		
170 pounds retail beef at $3.00/lb.	=	$510
130 pounds lean beef trimmings at $1.00/lb.	=	$130
130 pounds ground beef at $.50/lb.	=	$ 65
Total value of products		$861
Gross Profit per head		$ 36

A processor could now buy live cattle futures, and simultaneously sell lean beef trimmings and ground beef futures and spread hedge the positive processing margin. Likewise, if the processing margin were negative, the processor would sell live cattle futures and buy the beef products futures for a reverse spread hedge. Notice that the ground beef and lean beef trimmings account for only 23 percent of the total carcass value, thus this type of spread has a large risk factor that the margin is not protected. The retail beef (steaks and roasts) value is the major component, and is subject to price risk. Of course a packer could opt to change the amount of lean beef trimmings and ground beef by putting more retail cuts into the mix, but given the relative prices, that is generally not going to happen because it is not profitable to do so. Thus, this type of spread hedge for processors, although possible, should be entered into with caution and lots of information.

The most widespread use of the beef products futures is to hedge the total output of the processor for each product. If a processor produces 200,000 pounds of ground beef in a day, he could hedge that one product against decreasing prices and likewise for lean beef trimmings. By the same token, a food processor and/or supermarket that is in the market daily to buy ground beef and lean beef trimmings could hedge against the price of the product going up. Both of these situations are shown in Tables 8.6 and 8.7.

Table 8.6 Beef Processor Hedge

Cash	Futures
January 1	
Producing 200,000 pounds of lean beef trimmings (current price at $1.00/lb.)	Sell four lean beef contracts at $1.03
January 2	
Sell 200,000 pounds of lean beef trimmings at $.98/lb.	Buy four lean beef contracts at $1.01
	$ .02
Net Hedge Selling Price = $.98 + $.02 = $1.00/lb.	

Table 8.7 Food Processor Hedge

Cash	Futures
January 1	
Will need 100,000 pounds of ground beef in 2 days (current price $.50/lb.)	Buy two ground beef contracts at $.51/lb.
January 3	
Buy 100,000 pounds of ground beef at $.54/lb.	Sell two ground beef contracts at $.56/lb.
	$.05/lb.
Net Hedge Buying Price = $.54 – $.05 = $.49/lb.	

Dairy Products

Dairy processors, producers, and distributors gained a futures hedging opportunity with the introduction of cheddar cheese and nonfat dry milk futures and option contracts in June 1993 and later with new contracts for fluid milk and butter. Direct cheddar cheese short hedges will likely be placed by cheese producers or by distributors, **brokers**, packagers, and processors as inventory hedges. Direct long hedges may be implemented by cheese buyers, including the makers of soup and ready-to-eat foods, other processed food product manufacturers, restaurant chains and grocery chains, as well as distributors, brokers, and packagers. The cheddar cheese futures contract calls for delivery of 40,000 pounds of USDA Grade A or better cheddar cheese in 40-pound blocks. Both long and short hedges work in the same manner as any other processor hedge, as illustrated in the short hedge example of Table 8.8.

Table 8.8 Cheese Manufacturer Short Hedge

Cash	Futures
November 1	
Have cheese in inventory current price $1.40/lb.	Sell December futures at $1.55/lb.
December 1	
Sell product out of inventory at $1.20/lb.	Buy December futures at $1.35/lb.
	+$.20/lb.
Net Hedge Selling Price = $1.20 + $.20 = $1.40/lb.	

The nonfat dry milk futures contract will probably be used as a hedge most often by candy manufacturers and other food processors who include milk solids in breads, cakes, and other baking and confectionery products, or by nonfat dry milk brokers as a (short) inventory hedge. A long candy manufacturer hedge is illustrated in Table 8.9.

Table 8.9 Candy Manufacturer Long Hedge Using Nonfat Dry Milk Futures

Cash	Futures
March 1	
Forward sells candy based on today's dry milk price of $.30/lb. for delivery in 3 weeks.	Buys April futures at $.32/lb.
March 15	
Buys dry milk to process into candy at $.38/lb.	Sells April futures at $.40/lb.
–$.08/lb.	+$.08/lb.
Net Hedge Buying Price = $.38 – $.08 = $.30/lb.	

As fluid milk prices edge more and more toward open market price setting, dairy producers (as well as processors) find that they have a greater need to hedge to protect narrow profit margins. The new fluid milk contract allows producers to hedge their milk production. Cross hedging opportunities using cheddar cheese and nonfat dry milk futures now exist for fluid milk as well as for processors of other types of cheeses such as mozzarella. Consider also that milk processors can now hedge the cost of the raw produce (fluid milk) and several outputs (cheese, butter, and nonfat dry milk).

Input-Packer Hedges

Livestock become inputs in the final production process of producing meat and at various phases in the process of getting the livestock to market weights. Feeder cattle are an input in the production process of producing live (fed) cattle. Market hogs and live cattle are inputs in the production process of producing beef and pork for consumption.

For a live cattle producer, feeder cattle are the major input. The live cattle producer usually confines the feeder cattle in a feedyard and provides grains, supplemental protein, and roughage as feed, as well as other inputs such as credit and management. If the cattle producer has made a decision to produce fed cattle, then she must acquire the inputs. This cannot be done immediately, thus there exists the risk of a price increase in the inputs before they can be acquired. In order to maximize the use (and thus minimize per-unit fixed costs) of feeding facilities and associated equipment, the feeder may also have a constant marketing plan such that every two weeks a pen of cattle is placed on feed. This decision to buy feeder cattle every two weeks exposes the feeder to the risk of price increases between each purchase point. The feeder can protect against the risk of price increases by long hedging feeder cattle. Table 8.10 shows a feeder cattle input hedge.

Table 8.10 Feeder Cattle Input Hedge

Cash Position	Futures Position	Basis
May 1		
Anticipates putting cattle on feed in one week (feeder cattle price at $68.00/cwt.	Buy April feeder cattle contract at $69.00/cwt.	$1.00
May 8		
Buys feeder cattle and puts on feed at $69.50/cwt.	Sell April feeder cattle contract at $70.50/cwt.	$1.00
	+$ 1.50/cwt.	no change
Net Hedge Buying Price = $69.50 – $1.50 = $68.00/cwt.		

Meat packers transform livestock into consumer meat products and need market cattle and hogs as inputs into the meat production process. Most packers **forward sell** products based upon a certain price. Oftentimes the cattle or hogs have not been purchased, and thus the risk exists that prices will increase before the purchases are made. This risk can be protected using a long hedge on either live cattle or hogs. Tables 8.11 and 8.12 show packer hedges for live cattle and hogs.

Table 8.11 Live Cattle Packer Hedge

Cash Position	Futures Position	Basis
May 1		
Forward sell meat products; based upon today's live cattle price at $60/cwt.	Buy April feeder cattle futures at $62/cwt.	$2
May 6		
Buys fed cattle at	Sell April feeder cattle futures at	
$62	$65	$3
–$2	+$3	change of $1
Net Hedge Buying Price = $62 – $3 = $59/cwt.		

Table 8.12 Hog Packer Hedge

Cash Position	Futures Position	Basis
September 1		
Forward sell meat products; based upon today's hog price at $40/cwt.	Buy October hog contract at $42/cwt.	$2
September 8		
Buys hogs at	Sell October hog futures contract at	
$38/cwt.	$41/cwt.	$3
–$2/cwt.	–$1/cwt.	change of $1
Net Hedge Buying Price = $38 – (–$1) = $39/cwt.		

Merchandisers and Processors

Some cattle and hog traders purchase and resell animals trying to profit from the trades. They typically do not hold the animals for long periods of time (usually never over two weeks because of the price risk and because of feeding and handling problems). There is also a type of cattle trader called a backgrounder who buys cattle, puts them in a feedyard-type environment, and preconditions them either to be sold and put on pasture and/or placed in the feedyard. Preconditioning involves all types and ages of cattle. Pork processors buy market hogs and process them into retail cuts of pork, only one of which has a futures contract—pork bellies. Pork processors face the risks of market hog prices increasing and pork belly prices decreasing. Beef processors face the risks of cattle prices increasing and the price of finished beef (lean beef, lean beef trimmings, and ground beef) decreasing.

A large institutional user of shrimp faces the risk that the price of shrimp will increase. The institution could long hedge shrimp to protect purchases and would likely scale the hedge as illustrated in Table 8.13.

Table 8.13 Institutional Shrimp Hedge

Cash	Futures
January 1	
Will need 5,000 lbs. of white shrimp daily for three days beginning in one week (current price is $4.50/lb.)	Buys three February shrimp contracts at $4.75/lb.
January 8	
Buys 5,000 lbs. of white shrimp at $4.55/lb.	Sells one February contract at $4.80/lb.
January 9	
Buys 5,000 lbs. of white shrimp at $4.60/lb.	Sells one February contract at $4.84/lb.
January 10	
Buys 5,000 lbs. of white shrimp at $4.70/lb.	Sells one February contract at $4.90/lb. January 8 = $.02 January 9 = $.03 January 10 = $.06
Net Hedged Buying Price = [$4.55 – $.05] + [$4.60 – $.09] + [$4.70 – $.15] / 3 = $4.52/lb.	

Trade Hedges

Cattle traders and backgrounders have both increasing and decreasing price risk, depending upon which cash position they hold. Consider a trader who has purchased feeder cattle and plans to sell them as soon as possible, and thus faces price risk that while the long cash position is held, prices will decrease. A short hedge will provide protection, as shown in Table 8.14.

Table 8.14 Cattle Trader Short Hedge

Cash Position	Futures Position	Basis
November 1		
Buys feeder cattle at $60/cwt.	Sells December feeder cattle contract at $61.00/cwt.	$1.00
November 3		
Sells feeder cattle at $58.00/cwt. –$2.00/cwt.	Buys December feeder cattle contract at $59.50/cwt. +$1.50/cwt.	$1.50 change of $.50
Net Hedge Selling Price = $58.00 + $1.50 = $59.50/cwt.		

Cattle traders may also forward sell if they can find selling opportunities. If they forward sell and do not have a long cash position to deliver against the forward sell, then they have the risk that cash prices will increase. A long hedge can offer protection against such a price risk (illustrated in Table 8.15).

Table 8.15 Cattle Trader Long Hedge

Cash Position	Futures Position	Basis
November 1		
Forward sell feeder cattle at $61/cwt.	Buy December feeder cattle contract at $63/cwt.	$2
November 15		
Buy feeder cattle at $64/cwt.	Sell December feeder cattle contract at $67/cwt.	$3
– $3/cwt.	+ $4/cwt.	change of $1
Net Hedge Buying Price = $64 – $4 = $60/cwt.		

Value-Added Hedges

Processors who add value to raw meat through further processing have two major price risk scenarios: an initial forward cash sale with a corresponding long cash position entered into at some subsequent date, and an initial long cash position without a corresponding forward sale. The first scenario was discussed and illustrated in the Input-Packer Hedges section. The second scenario, a long cash position without a corresponding cash forward sale, exposes the processor to the risk that the output price will decrease.

If a pork processor, for example, has purchased market hogs and is processing them, his risk is that the retail prices will decrease and potential profits will be squeezed and/or eliminated. Ideally the processor would like to have protection for all the retail pork products produced, but futures exist only on pork bellies. The processor can hedge with pork belly futures to protect a portion of the retail cuts produced (Table 8.16). Care must be taken to balance the amounts of actual bellies produced with the futures position. Although only the pork bellies contract offers the hog processor a direct hedging device, a potential exists for cross hedging some or all the other retail cuts. Cattle processors can do likewise with the lean beef, lean beef trimmings, and ground beef futures contracts. These three contracts do not represent all the retail cuts of beef in a carcass but do provide adequate hedging opportunities.

Table 8.16 Pork Processor Hedge

Cash Position	Futures Position	Basis
September 15		
Have pork bellies in storage and unsold (current price of $78/cwt.)	Sell October pork belly futures contract at $80/cwt.	$2
September 20		
Sell pork bellies at $74/cwt.	Buy October pork belly contract at $76/cwt.	$2
	+$4/cwt.	no change
Net Hedge Selling Price = $74 + $4 = $78/cwt.		

Using Target Prices

Target prices are very helpful to livestock hedgers in evaluating whether to enter the production process and in deciding whether to market animals heavier or lighter. A cattle feeder can calculate a target price and estimated production costs to determine whether to put cattle on feed before the feeder cattle are purchased and the process begins. Likewise, a cattle trader can calculate a target price and production costs to determine whether or not to enter a long cash position. If the production process is already underway, then **target prices** can help producers evaluate the timing of the cash sales. Figure 8.1 provides examples of basis for feeder cattle with two different delivery months.

April Contract, Weekly Average, Feeder Cattle Basis Information, Choice 500–600 Pound Steers for Clovis, NM in dollars per hundredweight, 1972–1992

Date	Average	Median	High	Low
Nov. 1st week	2.2716	1.018	6.880	−1.360
2nd	2.4893	1.5675	6.4725	−1.6325
3rd	2.8844	2.5167	6.054	−1.890
4th	1.7029	1.6784	3.0975	0
Dec. 1st week	1.2910	1.441	3.880	−1.8388
2nd	2.2315	1.925	5.535	−.530
3rd	2.7038	2.742	5.418	−.540
4th	2.3597	3.130	4.155	−.206
Jan. 1st week	2.3988	1.900	9.5743	−2.750
2nd	2.0986	1.010	8.150	−1.100
3rd	1.7903	1.734	6.928	−1.055
4th	1.0244	.550	6.945	−4.478
Feb. 1st week	1.1786	.974	8.548	−4.005
2nd	.5914	.164	8.235	−4.412
3rd	.4434	.4333	6.180	−2.000
4th	.3243	.300	4.600	−3.320
Mar. 1st week	.2601	.214	4.220	−2.000
2nd	−.2056	.375	4.534	−6.790
3rd	.5630	.990	4.350	−4.100
4th	.2224	.035	4.260	−3.00
Apr. 1st week	−.4848	−.4357	6.784	−6.632
2nd	−.0249	.1438	9.488	−8.462
3rd				
4th				

Figure 8.1 Feeder cattle basis information—April

A cow-calf producer who normally sells feeders in the first week of January can compute a target price as follows:

Current futures quote for March feeder cattle	$71.00
Average basis 1st week of January (Clovis, NM)	$ 2.39

Target Price = $71.00 – $2.39 = $68.61 for the first week in January. If the operator waits until the fourth week in January, the target price is $71.00 – $1.02 = $69.98.

The operator can decide if the extra three weeks are cost-effective—that is, will the cost per hundredweight be less than the difference between the two target prices ($69.98 – $68.61 = $1.37)?

This adds more information so that the operator can decide to market the animals heavier or lighter.

To-Bid Prices

To-bid prices are used by some livestock traders and firms to determine cash quotes. The process involves a current futures quote for the nearby futures, a transportation differential, carrying charges, and profit. Livestock are storable only for very brief time periods because they either continue to grow and change product forms or they deteriorate. Carrying charges are therefore valid for very brief periods (a few days at most). Profit levels are determined principally by the amount of competition that exists in an area or region.

A live cattle buyer sees that the nearby futures are trading at $67. She figures a two dollar transportation charge but no carrying charge differentials exist and projects a need for a two dollar profit. She is therefore willing to pay $63 for fed cattle. If she buys cattle for $63 and then can't find a buyer, she could transport them to the nearest futures delivery point (at a cost of $2/cwt.), deliver against the futures, and receive $67, and thus a profit of $2/cwt. This approach works only if the cattle have no carrying charges (i.e., the current time must be during a maturing futures contract month). Otherwise, carrying charges must be considered. However, the time period must not be very long because cattle that are finished cannot be held for too long or they will not meet the conditions of the futures contract for grades and weights. This is true for all types of meat animals. Thus, to-bid prices for livestock are less precise than for storable commodities such as grains, metals, and some financial instruments.

Because livestock are considered to be nonstorable, some firms calculate to-bid prices by simply subtracting a "basis" factor. This basis factor is really just a catch-all value that includes personal risk preferences. They take the futures quote and subtract the basis factor and call it a to-bid price. This is surprisingly common in certain parts of the country, especially where competition is weak or does not exist at all.

Total Hedging

Cattle feeders use several inputs that have futures contracts such as grains, soybean meal, feeder cattle, and credit. Additionally the output, fed cattle, has a futures contract. Consequently some feeders hedge the output to protect against decreasing prices and hedge as many inputs as possible to protect against increasing prices. This process is called total hedging.

Total hedging requires a great deal of record keeping to make sure that the hedges are placed properly and timed when the hedges need to be placed. Many total hedging programs are continuous hedging programs, that is, the hedgers do not selectively hedge. They will hedge all their corn purchases, all their feeder cattle purchases, all their soybean meal purchases, and all their cash sales of fed cattle all the time. Some feeders might hedge only the grain inputs and selectively hedge the output. This situation is especially true for custom feeders who do not take title to the cattle placed on feed in their feedyards.

Basis Aspects

With storable commodities such as grains, the difference between a cash price and a futures price logically represents transportation charges and carrying charges. However, for nonstorable commodities such as livestock, the difference between a cash price and a futures price cannot be explained by transportation and carrying charges. Transportation charges are a logical component of livestock basis, and certainly a few days of carrying charges are valid; however, any remainder is not logically explainable except as representing expectations of future supply and demand conditions.

Because of the uncertainty of basis values in livestock, selective hedging based upon basis patterns is very difficult. Most livestock basis patterns show a gradual narrowing as the contract matures; however, the pattern does not always narrow without several movements up and down.

Also, as of January 1987, the feeder cattle contract has a cash settlement provision. Cash settlement means that delivery against the futures contract of actual feeder cattle is no longer possible. If a trader decides to deliver against the contract, a cash settlement is provided in lieu of the actual cattle. The cash settlement is based upon an index of actual feeder cattle prices from twenty-seven cash markets as reported and compiled by CATTLEFAX, an independent price reporting service located in Denver, Colorado. Basis for feeder cattle must now be calculated using the twenty-seven-market index reported by CATTLEFAX.

Some livestock products have storable components to them. Butter, cheese, and nonfat dry milk can be stored for several months without loss of product quality. Basis for these products will have a cost of carry component. Lean beef, lean beef trimmings, and ground beef are perishable and thus will have very little if any cost of carry, unless frozen, which would then render them unsuitable to deliver against the futures. Nonetheless, there is a large cash market in frozen beef products, so the cost of carry is not totally useless as a measure for some basis values, but it should not be the major factor in determining basis values.

PROBLEMS

1. CattleGrow, Inc. is a large feedyard that buys feeder cattle, feeds them for several weeks, and then sells them as fed cattle. They also buy large volumes of corn and soybeans to process into rations for the cattle. What are their major price risks and how would they construct a total hedging program?

2. Very Fast Burgers (VFB), Inc. is a large fast food chain that sells each day at least 100,000 pounds of ground beef. Every day they enter the market place and buy 100,000 pounds of ground beef. Today they paid $.49 per pound for the beef. The nearby ground beef futures is trading at $.52 per pound today. A supplier offers to sell them beef tomorrow for "four cents off the nearby." VFB takes the offer and immediately buys two nearby futures at the current price of $.52 per pound. Tomorrow, the nearby futures is at $.55 per pound, so the supplier sells the cash beef to VFB at a price of $.51 per pound and VFB lifts their hedge. What is VFB's net hedge buying price? Do you think that VFB did the right thing by taking the supplier's offer? Explain.

3. Seth is a cattle jobber. He buys feeder cattle and sells them within a few days, or if he finds someone who needs cattle, he will forward sell the cattle and then buy them to fulfill the order. How could Seth use the feeder cattle futures market?

4. What are some of the risks associated with a beef processor using live cattle futures and the beef products futures in a spread hedge?

CHAPTER 9

Hedging Strategies: Metals and International Commodities

"If you get up early, work late, and pay your taxes, you will get ahead—if you strike oil."

—J. Paul Getty

Cotton futures trading opened in New York in 1870, and sugar futures trading was initiated on the Coffee, Sugar, and Cocoa Exchange in 1941. Cotton and sugar, and in later years, gold, silver, platinum, palladium, coffee, and frozen concentrated orange juice, as well as the petroleum complex and several financial derivatives established New York as the U.S. trading center for commodities other than grains, oilseeds, and livestock. The financial futures contracts begun in the 1980s re-established Chicago as the futures trading center. But, in the meantime the commodities traded in New York reinforced the image of New York as an international trading center.

The petroleum contracts have proven to be both popular and controversial in the petroleum industry. The methods and mechanisms of hedging with any of these contracts is very similar to hedging with grains and oilseed futures because they are storable commodities.

Hedging the Metals Complex

Several metals, including copper, gold, silver, platinum, and palladium have futures contracts. Producers and processors (mining firms) and smelters, traders, and end users can use the metal futures to protect against adverse price movements. Producers and processors are concerned that prices will fall during the production and/or refining process. Also, if they forward sold the bulk processed commodity at some price, then they would be risking price increases. Traders, likewise, can protect inventories and forward sales. End users need protection against forward purchases and inventory changes.

Short Hedging

Consider a smelting operation for copper. The raw ore is purchased and refined into copper. During the smelting process, the firm has the risk that copper prices will fall. They have certain input costs such as the ore itself (usually based on refined copper prices), fixed costs (including their facilities), and other variable costs (such as electricity and labor). The risk the smelter faces is that finished copper prices will decrease after the raw ore purchase and eliminate or reduce profit margins or dip below the cost of production. A short hedge can provide protection against this price decline. The example in Table 9.1 shows both a price increase and decrease with no basis change. Table 9.2 shows a basis improvement and deterioration.

A similar hedge would be placed by a mining firm that has already either purchased the ore reserve or leased the mineral rights to a specific reserve at a fixed lease price per time period or per unit extracted. This hedge works particularly well if the raw ore is sold to a smelter at a price that is dependent upon the price of the refined commodity or if the raw ore is to be refined by the same firm that mines it.

Short hedges are also used by traders (dealers) who have the metal in inventory and need protection against a decrease in the cash price before the

Table 9.1 Short Hedge for a Metal Refinery

Cash	Futures	Basis
May 1		
Smelting operations in progress (current copper price is $.75/lb.)	Sell July copper futures at $.76/lb.	.01
	Price Decrease	
May 15		
Sell copper at $.73/lb.	Buy July copper futures at $.74/lb.	.01
opportunity loss .02	+.02	no change
	Net Hedged Price = $.73 + .02 = $.75/lb.	
	Price Increase	
May 15		
Sell copper at $.77/lb.	Buy July copper futures at $.78/lb.	.01
opportunity gain .02	– .02	no change
	Net Hedged Price = $.77 – .02 = $.75/lb.	

Table 9.2 Short Hedge for a Metal Refinery with Basis Changes

Cash	Futures	Basis
May 1		
Smelting operations in progress (current copper price is $.75/lb.).	Sell July copper futures at $.76/lb.	.01
	Basis Improvement	
May 15		
Sell copper at $.74/lb.	Buy July copper futures at $.74/lb.	.00
opportunity loss .01	+.02	change of .01
	Net Hedged Price = $.74 + .02 = $.76/lb.	
	Basis Deterioration	
May 15		
Sell copper at $.73/lb.	Buy July copper futures at $.75/lb.	.02
opportunity loss .02	– .01	change of .01
	Net Hedged Price = $.73 – .01 = $.72/lb.	

inventory is priced and sold. Consider a palladium dealer who has palladium purchased at $140 per ounce and is currently trying to find a buyer. The risk is that he cannot find a buyer at a price of $140 plus carrying costs and a profit margin. Thus a price decrease would eliminate any possibility of a profit or even cost recovery. A short hedge as illustrated in Table 9.3 will provide the protection that the dealer needs. A similar situation exists for end users who have metals in inventory and need protection against declining prices in the raw commodity that affect the prices of finished goods such as gold jewelry or copper wire.

Table 9.3 Short Hedge for a Metals Dealer

Cash	Futures	Basis
February 1		
Buys palladium at $140/oz.	Sells March palladium futures at $142/oz.	$2
	Price Decrease	
February 15		
Sell palladium at	Buys March palladium futures at	
$135/oz.	$137/oz.	$2
– $5/oz.	+ $5/oz.	no change
	Net Hedged Price = $135 + $5 = $140/oz.	
	Price Increase	
February 15		
Sell palladium at	Buys March palladium futures at	
$142/oz.	$144/oz.	$2
+ $2/oz.	– $2/oz.	no change
	Net Hedged Price = $142 – $2 = $140/oz.	

Long Hedging

Long hedging metals is usually a process that involves a short cash position such as a forward sale/anticipated purchase. Refining operations, dealer-traders, and end users often have the opportunity to forward sell products before they actually take possession of the products. They anticipate purchasing the product at some point in the future and then tendering delivery against the forward sale before the expiration of the deadline. They face the risk that if they forward sell at a specific price when they go into the cash market to actually purchase the product for delivery tender, the price has increased and eliminated profit and/or caused a loss. A long hedge, as illustrated by the following example, can offer protection.

A jewelry manufacturer has an opportunity to forward sell some gold jewelry to a wholesaler for delivery in one month. He must price the jewelry to the wholesaler today. He does so based upon the current price of gold at $450 per ounce. The manufacturer faces the risk that, between now and when he buys the gold to manufacture into jewelry, the price will be above $450 per ounce. A long hedge will protect against increasing prices. This particular manufacturer will probably structure a scale-down hedge because the contract was large and not all the gold will be purchased at once, as illustrated in Table 9.4.

Table 9.4 Long Hedge for a Jewelry Manufacturer

Cash	Futures
May 1	
Sells Jewelry based upon current price of gold at $450/oz.	Buys three August Gold futures at $470/oz.
May 15	
Buys 100 oz. of gold at $455/oz.	Sells one August Gold futures at $475/oz. + $5/oz. for one contract
June 1	
Buys 100 oz. of gold at $460/oz.	Sells one August Gold futures at $480/oz. + $10/oz. for one contract
June 15	
Buys 100 oz. of gold at $475/oz.	Sells one August Gold futures at $495/oz.
Losses: $ 5/oz. 100 oz. $10/oz. 100 oz. $25/oz. 100 oz.	Gains: $ 5/oz. 100 oz. $10/oz. 100 oz. $25/oz. 100 oz.
Net Hedged Price = $455 – $ 5 = $450/oz. $460 – $10 = $450/oz. $475 – $25 = $450/oz.	

A dealer might be considering buying cash gold at some point in the future and need protection against increasing prices. Likewise, an investor might have decided to invest in gold in two weeks when her Treasury bills have been liquidated and converted to cash. Both of these anticipated purchases could be hedged using a long hedge, as illustrated in Table 9.5.

Table 9.5 Anticipatory Long Hedge for a Metals Dealer

Cash	Futures
July 15	
Anticipates in two weeks gold will be purchased (current price is $440/oz.)	Buy one August gold contract at $450/oz.
Price Increase	
August 1	
Buys gold at $450/oz.	Sell one August gold contract at $460/oz.
	+$ 10/oz.
Net Hedged Price = $450 – $10 = $440/oz.	
Price Decrease	
August 1	
Buys gold at $430/oz.	Sell one August gold contract at $440/oz.
	–$ 10/oz.
Net Hedged Price = $430 – (–$10) = $440/oz.	

Spread Hedging

Protecting a margin is sometimes more important than protecting against absolute price movements of a particular commodity, especially if that margin is based on a relationship between two or more commodities. Let's consider a smelting/refining operation that manufactures an alloy that has a fixed proportion of gold and silver. They operate in a competitive market and margins are very thin. They do not change the proportion of gold and silver in the alloy as prices change because the alloy must maintain its fixed proportions or cease to be the product that buyers want. This manufacturer not only needs protection against the price of metals moving up but also against one moving up more than the other and changing the very thin profit margins. A spread hedge will protect both against adverse absolute price movements and relative price movements. The manufacturer uses the gold/silver ratio as a pricing guide. The current price of gold is at $450 and silver at $8 per ounce for a gold/silver (GS) ratio of 56.25. The manufacturer uses this in a pricing formula for the final alloy product. When the GS ratio changes from 56.25, the price of the final product will change and/or the manufacturer's profit margins are affected. Hedging the GS ratio is possible with a spread hedge on gold and silver futures. If the ratio drops below the level the manufacturer sets (i.e., 56.25), then the manufacturer will buy gold futures and sell silver futures. There are some important considerations with this type of hedge. First, it is a very weak spread. There is no good evidence to suggest that gold and silver should be trading at some ratio over a given time period. Therefore, arbitragers may or may not cause the markets to keep a certain ratio. Secondly, because there is not a good economically sound basis for the spread, this type of spread hedge can be very risky. If a firm wants to spread hedge metals, they should have some good empirical evidence that the spread or ratio maintains itself over some time period, otherwise a spread hedge potentially may be more risky than maintaining a speculative cash position. Table 9.6 shows a gold/silver spread hedge, and Table 9.7 shows a gold/silver spread hedge that didn't work.

Table 9.6 Successful Spread Hedge for a Metals Refinery

Current Price for July Gold = $450.00/oz.
Current Price for July Silver = $8.00/oz.
Current GS Ratio = 56.25

July Gold increases to $453/oz.
July Silver remains unchanged
GS Ratio = 56.63

Spread hedge placed by:
Selling July Gold = $453.00/oz.
Buying July Silver = $8.00/oz.
The number of contracts of each would depend on the alloy mix.

July Gold decreases to $452.00/oz.
July Silver increases to $8.04/oz.

Spread lifted
Buy July Gold at $452.00/oz.
Sell July Silver at $8.04/oz.

July Gold = + $1.00/oz.
July Silver = + $ .04/oz.
Net = $1.04/oz.

Final GS Ratio 56.22 which is closer to the beginning spread of 56.25 than the before spread hedge spread of 56.63. In other words, the spread hedge protected part of the spread risk but not all. The GS ratio of 56.25 says that when silver prices move 1 cent, the price of gold must move 56.25 cents. In the above example when silver prices moved 4 cents, gold prices changed by 100 cents instead of the perfect spread movement of 225 cents (56.25 × .04).

Table 9.7 Unsuccessful Spread Hedge for a Metals Refinery

Current Price for July Gold	=	$450/oz.
Current Price for July Silver	=	$8/oz.
Current GS Ratio	=	56.25

July Gold increases to $453/oz.
July Silver remains unchanged
GS Ratio = 56.63

Spread hedge placed by:

Selling July Gold	=	$453/oz.
Buying July Silver	=	$8/oz.

July Gold increases to $475/oz.
July Silver remains unchanged

Spread lifted
Buy July Gold at $475/oz.
Sell July Silver at $8/oz.

July Gold = – $22/oz.
Net = – $22/oz.
July Silver = 0

Final GS Ratio 59.40 and the spread hedge did not protect the price relationship.

Basis Aspects

Basis for metals is conceptually similar to basis for grains. The theoretical reason for a metals basis is the carry costs–the difference between a cash price today and a futures price should be the cost of carrying the commodity to the future time period. As time passes and the life of the futures contracts expires, the cash and futures prices should converge during the maturity month.

Economic Factors

Precious metals have price movements that are generally demand driven. Their price is determined by supply and demand, but because of the miner's need to continue production as evenly as possible to spread fixed costs, most price changes are a result of changes in demand. The change in demand for precision metals comes primarily from changes in the tastes and preferences of consumers.

Precious metals have utility because of their metallurgical properties and because they are also perceived as a store of value. The metallurgical demand for metals is fairly constant. However, individual's attitudes about metals as a store of value changes with factors such as inflation, general economic conditions, wars, threats of war, and other expectations of the future.

When gold prices moved from below $400 per ounce to over $800 per ounce during the late 1970s and early 1980s and then collapsed to below $400, it was because of changes in people's tastes and preference concerning gold as a store of value. Inflation and general economic conditions are easier to predict than factors such as war and political unrest. Thus, forecasting precious metal prices is difficult at best. Because metals prices can move very rapidly and move significant amounts, and because profit margins are generally very thin in all of the metals and depend upon large volumes, hedging is crucial.

Hedging the Petroleum Complex

Energy futures began with a futures contract on heating oil in 1978. In the 1980s, several others were added, such as unleaded gasoline, propane, and crude oil. In April 1990 a natural gas contract was added. Each of these contracts can be used to hedge different cash positions. Jobbers (wholesalers) can use futures contracts to protect against anticipated future purchases and inventory changes of heating oil, propane, and unleaded gasoline. Crude oil jobbers can do likewise as well as crude oil shippers and producers. Owners of oil wells can protect against the value of the owned or leased resource declining (a typical inventory problem). As with metals, there are three types of hedges: short, long, and a spread.

Short Hedging

Consider an inventory hedge for an owner of an oil well. The resource is in the ground and subject to several conditions before it is removed. These conditions include state and local regulations on pumping, environmental restrictions, and the management decisions of the owner of the lease. Recent history has shown the importance of an inventory hedge for owners of oil wells or the lease to pump the well. Royalties are most often expressed as a percentage of the well head price and thus vary directly with the price—the higher the value of the lease. Thus, both the owner of the mineral rights and the owner of the lease to pump the oil will see their asset decline in value as crude oil prices decrease and vice versa.

A simple short inventory hedge will protect against the declining value of the asset. If a mineral right owner has proven reserves of 100,000 barrels and wants protection against declining crude oil prices, she could protect the inventory with a hedge by selling 100 futures contracts, as illustrated in Table 9.8.

Table 9.8 Petroleum Inventory Hedge

Cash	Futures	Basis
January 1		
Inventory of 100,000 barrels of oil (current price $22.00/barrel)	Sell 100 December crude oil futures at $22.07/barrel	$.07
	Price Decrease	
December 1		
Price decreases to $20.00/barrel	Buy 100 December crude oil futures at	
	$20.30/barrel	$.30
	+ $1.77	change of .23
Net Hedged Inventory Value = $20.00 + $1.77 = $21.77/barrel (Basis Deterioration of $.23/barrel)		
	Price Increase	
December 1		
Price increases to $23.00/barrel	Buy 100 December crude oil futures at	
	$23.05/barrel	$.05
	– $.98	change of $.02
Net Hedged Inventory Value = $23.00 – $.98 = $22.02/barrel (Basis Improvement of $.02/barrel)		

A jobber who has 100,000 barrels of gasoline in inventory has the risk that the price will fall before the gasoline is delivered to retail or other wholesale outlets and sold. The gasoline in inventory could be protected with the sale of 100 gasoline futures contracts. This is shown in Table 9.9.

Table 9.9 Gasoline Inventory Hedge

Cash	Futures	Basis
February 1		
Jobber has 100,000 barrels of gasoline in inventory (purchased at $.60/gal.)	Sells 100 March gasoline futures at $.63/gal.	$.03
	Price Decrease	
March 1		
Sells gasoline at	Buys 100 March gasoline futures at	
$.58/gal.	$.61/gal.	$.03
– $.02	+ $.02	no change
Net Hedged Price = $.58 + $.02 = $.60/gal.		

Table 9.9 Gasoline Inventory Hedge *(concluded)*

Cash	Futures	Basis
	Price Increase	
March 1		
Sells gasoline at	Buys 100 March gasoline futures at	
$.62/gal.	$.65/gal.	.03
+ $.02	– $.02	no change
Net Hedged Price = $.62 – $.02 = $.60/gal.		

Long Hedging

A heating oil jobber has found a wholesaler who will purchase 10,000 barrels of heating oil for delivery in two weeks. The forward sale is made and a price agreed upon of $.60 per gallon or $25.20 per barrel (42 gallons per barrel). The jobber must now find a refinery that he can purchase the heating oil from for less than $.60 per gallon. The jobber is concerned that prices will increase before the purchase can be made and the product delivered. The jobber could hedge by buying ten heating oil futures, as shown in Table 9.10.

Table 9.10 Long Hedge for Heating Oil

Cash	Futures
February 1	
Sells heating oil at $.60/gal. for delivery in two weeks	Buys 10 March heating oil contracts at $.61
Price Increase	
February 15	
Buys heating oil at	Sells 10 March heating oil contracts at
$.61/gal.	$.62/gal.
– $.01	+ $.01
Net Hedged Price = $.61 – $.01 = $.60/gal.	
Price Decrease	
February 15	
Buys heating oil at	Sells 10 March heating oil contracts at
$.58/gal.	$.59/gal.
+ $.02	– $.02
Net Hedged Price = $.58 + $.02 = $.60/gal.	

The same type of hedge could be used for refineries that are purchasing crude oil or jobbers who are purchasing gasoline and propane. Assume that a refinery purchases 100,000 barrels of crude oil every five days. Thus, they anticipate purchasing 100,000 barrels every five days and run the risk that the price will increase. They can hedge long, as illustrated in Table 9.11. If prices do increase, they are protected with the hedge, and if they decrease, the gain in the cash is offset by losses in the futures but they did not suffer a price increase.

Table 9.11 Long Hedges for a Refinery

Cash	Futures
February 1	
Anticipates purchasing crude oil every five days (100,000 barrels) (current price = $20.00/barrel)	Buys 100 March crude oil contracts at $20.50/barrel
February 6	
Buys 100,000 barrels at $21.00/barrel	Sells 100 march crude oil contracts at $21.50/barrel
	+ $1.00
	Buys 100 March crude oil contracts at $21.50/barrel
February 11	
Buys 100,000 barrels at $20.00/barrel	Sells 100 March crude oil contracts at $20.50/barrel
	– $1.00
Net Hedged Price = $21.00 – $1.00 = $20.00 (Feb. 1–Feb. 6)	
Net Hedged Price = $20.00 + $1.00 = $21.00 (Feb. 6–Feb. 11)	

Spread Hedging

Spread hedging with oil products is conceptually more appealing and easier to justify economically than with metals. There is a fundamental physical relationship between crude oil, heating oil, gasoline, and propane prices. Crude oil is the raw product, and heating oil, gasoline, and propane are all products of the cracking process. Thus, there is a physical yield relationship between and among these products that in turn can be translated to an economic relationship.

This relationship is much more similar to the soybean, soybean oil—soybean meal crush spreads and relationships discussed in Chapter 6 than it is to the gold-silver price ratio discussed earlier in this chapter. Because a physical/economic relationship exists in the petroleum cracking process, arbitragers (generally the day traders or pit scalpers) can bid away differences and profit from the changes in the normal relationships. Similarly, hedgers can take advantage of differences and spread hedge for protection.

If a refinery is primarily producing gasoline, then they are concerned with the margin between crude oil and gasoline. If a 42-gallon barrel of crude oil yields 17 gallons of gasoline, then they have a normal relationship that the price of the raw product plus **cracking** cost should equal the value of the product. If the value of the products exceeds the cost of crude plus the cost of cracking, then the refinery can place a cracking spread hedge. If the value of the products is less than the cost of crude and the cost of cracking, then the refinery can place a reverse cracking spread hedge.

Consider that the normal yield of cracking is 17 gallons of gasoline, 10 gallons of heating oil, and 15 gallons of other products and waste. The 15 gallons of other products (such as wax, asphalt, and naphtha) and waste is valued at $.40 per gallon for a total per-barrel value of $6.00. July crude is trading at $22 per barrel (42 gallons), July gasoline is trading at $.80 per gallon, and July heating oil is trading at $.60 per gallon. The values summarize as follows:

	Per Barrel of Crude Oil
July crude	$22.00
Cost of cracking	2.00
Total cost	$24.00
July gasoline (.80 × 17)	$13.60
July heating oil (.60 × 10)	6.00
Other products (.40 × 15)	6.00
Total value	$25.60
Difference (Profit)	+$ 1.60

Because the difference is positive, the refinery can place a cracking spread hedge by buying July crude oil futures and selling July heating oil futures and July gasoline futures. Because of the positive difference, the refinery can spread hedge and capture the difference as well as speculators that spread and other arbitragers. This action bids the price of crude up and the price of gasoline and heating oil down, and the difference approaches zero. The spread hedge is lifted, and the refinery captures a higher profit margin (Table 9.12).

Table 9.12 Spread Hedge for a Refinery

July crude at $22.00/barrel
July gasoline at $.80/barrel
July heating oil at $.60/gallon
Cost of cracking at $2.00/barrel
Value of other products $.40/gallon
Cracking yield: 17 gallons gasoline + 10 gallons heating oil + 15 gallons other = 42 gallons per barrel crude oil

Value of cracking:

$.80 × 17	= $13.60
$.60 × 10	= + 6.00
$.40 × 15	= + 6.00
Total value	= $25.60
Less cost of crude	= −22.00
Less cracking cost	= − 2.00
Net	+ $ 1.60/gallon

Put on Spread:
Buy July crude at $22.00
Sell July gasoline at $.80
Sell July heating oil at $.60

Lift Spread:
Sell July crude at $23.00
Buy July gasoline at $.78
Buy July heating oil at $.57

Net $1.00 July crude oil + $.34 (17 × $.02) July gasoline + $.30 (10 × $.03) July heating oil = $1.64

If the value of the product was less than $24.00 per barrel, then the refinery would sell crude oil and buy gasoline and heating oil futures, as illustrated in Table 9.13. This reverse spread is more risky than the normal cracking spread because the refinery cannot go through delivery if the markets don't normalize. In the example of the cracking spread, if the markets didn't go back to normal, the refinery can take delivery of the crude, crack it, and deliver against the gasoline and heating oil futures and make the difference. However, if the markets don't normalize with the reverse cracking spread, then the refinery cannot go through delivery and make the difference because it is short the crude and long the products (the refinery would have to deliver crude and accept delivery of the products). This action might be profitable, but logistical difficulties associated with this delivery process would likely indicate that it would not be a desirable activity.

Table 9.13 Reverse Spread Hedge for a Refinery

July crude at $22.00/barrel
July gasoline at $.75/barrel
July heating oil at $.55/gallon
Cost of cracking at $2.00/barrel
Value of other products $.35/gallon
Cracking yield: 17 gallons gasoline + 10 gallons heating oil + 15 gallons other = 42 gallons per barrel crude oil

Value of cracking:		
$.75 × 17	=	$12.75
$.55 × 10	=	+ 5.50
$.35 × 15	=	+ 5.25
Total value	=	$23.50
Less cost of crude	=	–22.00
Less cracking cost	=	– 2.00
Net		– $ .50/gallon

Put on Reverse Spread:
Sell July crude at $22.00
Buy July gasoline at $.75
Buy July heating oil at $.55

Lift Spread:
Buy July crude at $21.75
Sell July gasoline at $.75
Sell July heating oil at $.57

Net $.25 July crude oil + $.17 (17 × $.01) July gasoline + $.20 (10 × $.02) July heating oil = $.62

Natural Gas Hedging

The natural gas industry has changed dramatically during the last decade. The highly regulated, price stable 1960s and 1970s have been replaced with the relatively unregulated price volatile 1980s and 1990s. The Natural Gas Wellhead Decontrol Act of 1989 provided for complete decontrol of natural gas prices by January 1, 1993. Additionally, the structure of the market participants changed radically beginning in 1987 when the Federal Energy Regulatory Commission (FERC) issued orders that created open access transportation of interstate pipelines. This action created natural gas marketers and allowed producers to sell directly to end users rather than just to pipelines.

What has emerged during the last five years has been a very active spot market for natural gas. Long-term contracts are not being entered into because of the price volatility. Even short-term contracts have become somewhat unpopular because of price instability. Thus producers, pipeline companies, marketers, local distribution companies (LDCs), and end users have all been affected by the new price volatility. Producers are fearful of long-term contracts that would cause them to receive a lower price than the spot market. Likewise end users would love a long-term contract with a relatively low price, yet no one is willing to enter into a long-term agreement at a low price and then have to buy the gas on the spot market at higher prices. Price risk has become the major risk in the natural gas market.

Short Hedging

A natural gas well owner decides to bring a new well into production and figures a break-even price for gas of $1.25/MMBtus (one million British thermal units) and would like a profit margin of $.10/MMBtus. She observes that the current spot price is $1.45/MMBtus, thus well within her objective. However, by the time the well is into production, the price has fallen to $1.05/MMBtus and she faces not only no profit but a loss of $.20/MMBtus. To offset this risk, she could have hedged by selling a futures contract, as illustrated in Table 9.14.

Table 9.14 Natural Gas Well Owner Hedge

Cash	Futures
Developing a new well current price $1.45/MMBtus.	Sell one December natural gas contract at $1.65/MMBtus.
Price Decrease	
Gas flows from new well and is sold at $1.05/MMBtus.	Buy one December natural gas contract at $1.25/MMBtus.
	+$.40/MMBtus.
Net Hedged Selling Price = $1.05 + $.40 = $1.45/MMBtus.	
Price Increase	
Gas flows from new well and is sold at $1.70/MMBtus.	Buy one December natural gas contract at $1.90/MMBtus.
	– $.25/MMBtus.
Net Hedged Price = $1.70 – $.25 = $1.45/MMBtus.	

Notice in the example in Table 9.14 that whether prices went up or down, the producer received $1.45/MMBtus, which was above the target price of $1.35/MMBtus. If the producer had not hedged the gas, then when the price fell to $1.05/MMBtus, the shortfall would have been $.30/MMBtus from the target price. However, when the price went up to $1.70/MMBtus, the gain over the target price was $.35/MMBtus. This example points out the major reason that hedging is unpopular—favorable price movements are eliminated to gain protection for unfavorable price movements. When the producer hedged the price received regardless of price, movement was $1.45/MMBtus, but unhedged the potential price was either $1.05 or $1.70 per MMBtus. This example also points out the important advantage that hedging provides—the ability to control price risk. The producer set a profit goal of $.10/MMBtus and a target price of $1.35/MMBtus. Without hedging, the producer had no price protection and thus could have faced a losing or very profitable situation.

Long Hedging

If a natural gas marketing firm forward sells gas for delivery in one month at a price of $1.40/MMBtus but has not established the buying price of the gas, then the firm has the risk that prices will increase before the gas is purchased. Table 9.15 shows an example with prices both increasing and decreasing. The example is similar to the short hedge example. Regardless of whether prices increase or decrease, both long and short hedgers are protected from the risk of price changes.

Table 9.15 Natural Gas Marketer Hedge

Cash	Futures
Forward sells gas for delivery in one month at $1.40/MMBtus.	Buy one December natural gas contract at $1.70/MMBtus.
Price Increase	
Buys gas at	Sell one December natural gas contract at
$1.60/MMBtus.	$1.90/MMBtus.
–$.20/MMBtus.	+$.20/MMBtus.
Net Hedged Buying Price = $1.60 – $.20 = $1.40/MMBtus.	
Price Decrease	
Buys gas at	Sell one December natural gas contract at
$1.20/MMBtus.	$1.50/MMBtus.
+$.20/MMBtus.	–$.20/MMBtus.
Net Hedged Buying Price = $1.20 – (–$.20) = $1.40/MMBtus.	

Basis Aspects

Basis for petroleum products follows the familiar cost-of-carry concept. In addition, certain products must have an allowance for product changes over time. Gasoline can be stored, but not for extended periods of time if it is to maintain product quality. Thus, far away basis for petroleum products must have a built-in allowance for quality differences.

Economic Factors

Oil is very much an international commodity, and its price and trade are greatly influenced by world political situations. The last two decades have seen dominating governmental influences in the pricing of oil as well as ownership of reserves through nationalization. The formation of the group called OPEC (Organization of Petroleum Exporting Countries) caused major changes in the pricing of oil worldwide. As the influence of OPEC rises and falls, so does the impact on worldwide oil prices. Any serious student of oil prices must be an "OPEC Watcher."

The most fundamental economic theory concerning oil is the theory of the mine. Simply stated, the theory says that the rate of extraction of a nonrenewable resource is dependent upon the market rate of interest and expected price movements of the resource. If the expected price increase for the resource is 6 percent and the current market rate is 11 percent, then the owner will extract the resource at as high a rate as possible, sell the product, and invest at 11 percent. This action has a tendency to dampen the resource's price today. However, the more of a nonrenewable resource that is extracted today, the less that is available for the future, and thus price expectations for the future increase. At the same time, if the expectation for price increase is 11 percent and the current market interest rate is 6 percent, then owners will limit production currently, which drives prices upward and depresses future price expectations because more will be available to mine in the future. Thus the theory of the mine says that the price of the resource will increase at a rate equal to the current market interest rate—any deviations will be bid away as the markets for the product and for interest rates return to equilibrium. Of course, the theory describes the mechanism that a competitive market uses to return to equilibrium and thus must be subject to modifications to reflect another market structure such as the existence of a cartel.

A brief review of price performance over the past few years shows that the theory of the mine is fairly accurate even with the existence of a cartel group (OPEC) that keeps the price higher than it would be in a competitive market through supply management (a restriction in production mainly through quotas). Because current production is limited due to supply management, more will be available in the future. Therefore, price expectations for the future are decreased. At the same time, the higher than competitive price for oil caused more exploration for oil in non-OPEC regions (such as Alaska and the North Sea) and new reserves to be found. As the exploration and discovery process continued, and as individual OPEC member needs changed, OPEC's group power began to erode and supply became more difficult to control. The market settled into a rate of extraction more in line with what the theory of the mine predicts.

Other International Commodities

Futures markets are constantly changing by adding new contracts. Often thin markets will have their contract dropped. In 1986 and 1987, new futures contracts on high fructose corn syrup and on an index of several futures prices were added and then dropped due to lack of trading volume. There are several contracts that have existed for years that don't fall into the categories of grains, oilseeds, livestock, financials, metals, or petroleum. Commodities such as cocoa, coffee, orange juice, and cotton have successful futures contracts, but they don't fall into the broad categories mentioned previously. Cocoa, coffee, cotton, and orange juice are fairly stable commodities (in terms of both quantity and price) but are influenced by world conditions. They are storable commodities that are influenced by weather conditions. Hedging opportunities exist through both short and long hedging positions but spread hedging is not a viable alternative.

Producers, traders, processors, and wholesalers can hedge to protect inventories and forward purchases and sales. A coffee producer can protect a forward sale, and a cocoa processor can protect a forward purchase.

Short Hedging

An orange juice trader has purchased 30,000 pounds of juice for $1.25 per pound. The trader wants to sell the product as quickly as possible, but while it is inventoried he runs the risk that the price will decrease. A hedge can be placed by selling two contracts of July orange juice futures, as shown in Table 9.16.

Table 9.16 Orange Juice Inventory Hedge

Cash	Futures	Basis
March 1		
30,000 pounds of orange juice purchased at $1.25/lb.	Sell two March orange juice contracts at $1.28/lb.	.03
	Price Decrease	
March 5		
Sell at $1.22/lb.	Buy at $1.25/lb.	.03
– $.03	+ $.03	no change
	Net Hedged Price = $1.22 + $.03 = $1.25/lb.	

(continued on the following page)

Table 9.16 Orange Juice Inventory Hedge *(concluded)*

Cash		Futures		Basis
		Price Increase		
March 5				
Sell at	$1.28/lb.	Buy at	$1.31/lb.	.03
	+ $.03		– $.03	no change
Net Hedged Price = $1.28 – $.03 = $1.25/lb.				

A cotton producer has cotton planted and will harvest and sell in November. It's May 15, and the producer is worried that cotton prices will decrease. The producer can hedge by selling December cotton futures, as illustrated in Table 9.17. Additionally the cotton producer can calculate a target price using historical basis information such as in Figure 9.1. For example, if the producer plans to harvest and sell the first week of November, the basis for that week for a December contract is $4.55 per hundredweight on average for Phoenix, Arizona. If the current (May 15) price for December futures is $67.70, then the target price is $63.15 per hundredweight. The producer could make a decision about hedging, and if the target price is calculated before planting, then a decision concerning whether or not to produce can be made.

Table 9.17 Cotton Producer Hedge

Cash	Futures	Basis
March 15		
Cotton planted (current price at $.58/lb.)	Sell December cotton futures at $.60/lb.	.02
November 15		
Harvest and sell at $.48/lb.	Buy December cotton futures at $.50/lb.	.02
	+ $.10	no change
Net Hedged Price = $.48 + $.10 = $.58/lb.		

Long Hedging

A coffee processor has entered into a forward contract to deliver roasted coffee to a wholesaler for delivery in one month. He determines the forward contract price based upon today's raw coffee price of $1.10 per pound. If prices increase before the processor can buy the green coffee, process, roast, and deliver it, then the processor's profit margin shrinks and may become negative. The processor could hedge by buying coffee futures and be protected as shown in Table 9.18. The example shows both a price increase and a price decrease.

Date (Month/Week)	December Contract		
	Average	High	Low
Jan 1st	3.84	13.34	–1.53
2nd	2.05	14.80	–13.89
3rd	2.08	12.93	–13.46
4th	2.39	14.26	–8.21
Feb 1st	2.23	15.50	–7.69
2nd	2.81	15.75	–5.26
3rd	2.87	15.25	–3.59
4th	2.84	14.57	–4.18
Mar 1st	3.06	13.90	–4.42
2nd	3.61	12.51	–4.21
3rd	3.29	11.64	–5.54
4th	3.26	11.80	–5.87
Apr 1st	3.13	12.17	–4.44
2nd	2.59	11.92	–5.13
3rd	1.77	11.83	–7.13
4th	2.47	12.43	–6.42
May 1st	2.07	9.63	–6.62
2nd	2.14	9.67	–6.34
3rd	2.10	8.05	–6.71
4th	2.39	8.15	–4.36
Jun 1st	2.83	8.41	–1.60
2nd	2.81	7.30	–1.91
3rd	3.33	10.00	–2.54
4th	3.57	10.44	–1.32
Jul 1st	3.46	9.42	–0.90
2nd	3.91	8.90	–0.19
3rd	4.13	8.35	–0.86
4th	4.14	8.34	–0.49
Aug 1st	4.39	7.94	1.66
2nd	4.13	9.15	1.33
3rd	4.32	8.01	2.75
4th	4.12	6.85	1.78
Sep 1st	4.03	6.96	0.45
2nd	4.10	5.93	0.30
3rd	4.52	6.74	0.94
4th	5.24	7.32	3.00
Oct 1st	4.77	7.05	2.00
2nd	5.06	7.67	1.89
3rd	4.72	7.76	2.50
4th	4.31	6.84	2.77
Nov 1st	4.55	7.02	3.00
2nd	4.47	7.47	2.61
3rd	2.61	7.58	–9.37
4th	3.73	7.80	–0.18
Dec 1st	3.00	7.91	–2.15
2nd	2.15	12.15	–16.26
3rd	2.88	14.15	–17.42
4th	1.94	13.98	–21.82

Figure 9.1 Weekly cotton basis information for the Phoenix market,1973–1992

Table 9.18 Coffee Processor Hedge

Cash	Futures	Basis
May 1		
Sells roaster coffee based upon raw coffee price of $1.10/lb. for delivery in one month.	Buys July coffee futures at $1.15/lb.	.05
	Price Increase	
May 28		
Buys coffee to process at	Sells July coffee futures at	
$1.12/lb.	$1.17/lb.	.05
– $.02/lb.	+ $.02/lb.	no change
	Net Hedged Price = $1.12 – $.02 = $1.10/lb.	
	Price Decrease	
May 28		
Buys coffee to process at	Sells July coffee futures at	
$1.09/lb.	$1.14/lb.	.05
+ $.01/lb.	– $.01/lb.	no change
	Net Hedged Price = $1.09 + $.01 = $1.10/lb.	

Basis Aspects

Cocoa, coffee, cotton, and orange juice are storable commodities, and basis is theoretically the cost of carry. Figure 9.2 shows a cotton basis chart that shows the gradual narrowing of the basis as maturity approaches. Notice that during the delivery month the basis becomes very erratic as profit takers are arbitraging the markets.

Because these commodities' bases are strongly influenced by the cost of carry, basis trading is very prevalent. Traders and dealers can basis trade and operate on very thin margins. For example, a coffee trader may offer to buy coffee "two cents off," meaning that he is willing to pay two cents under the nearby futures contract. If the trader can negotiate this purchase, then he can immediately try to sell it at say one cent under the nearby. If he can get both trades established and hedged, he is assured of getting revenue of $.01 per pound. This example is illustrated in Table 9.18. The trader is unconcerned about price movements and levels. He will make $.01 per pound regardless of whether or not prices increase or decrease, and whether price levels are at $1.00 per pound or $20.00 per pound.

Economic Factors

All four of these commodities are supply driven. That is, most price movements of any size are caused by changes in supply, usually caused by weather. Coffee,

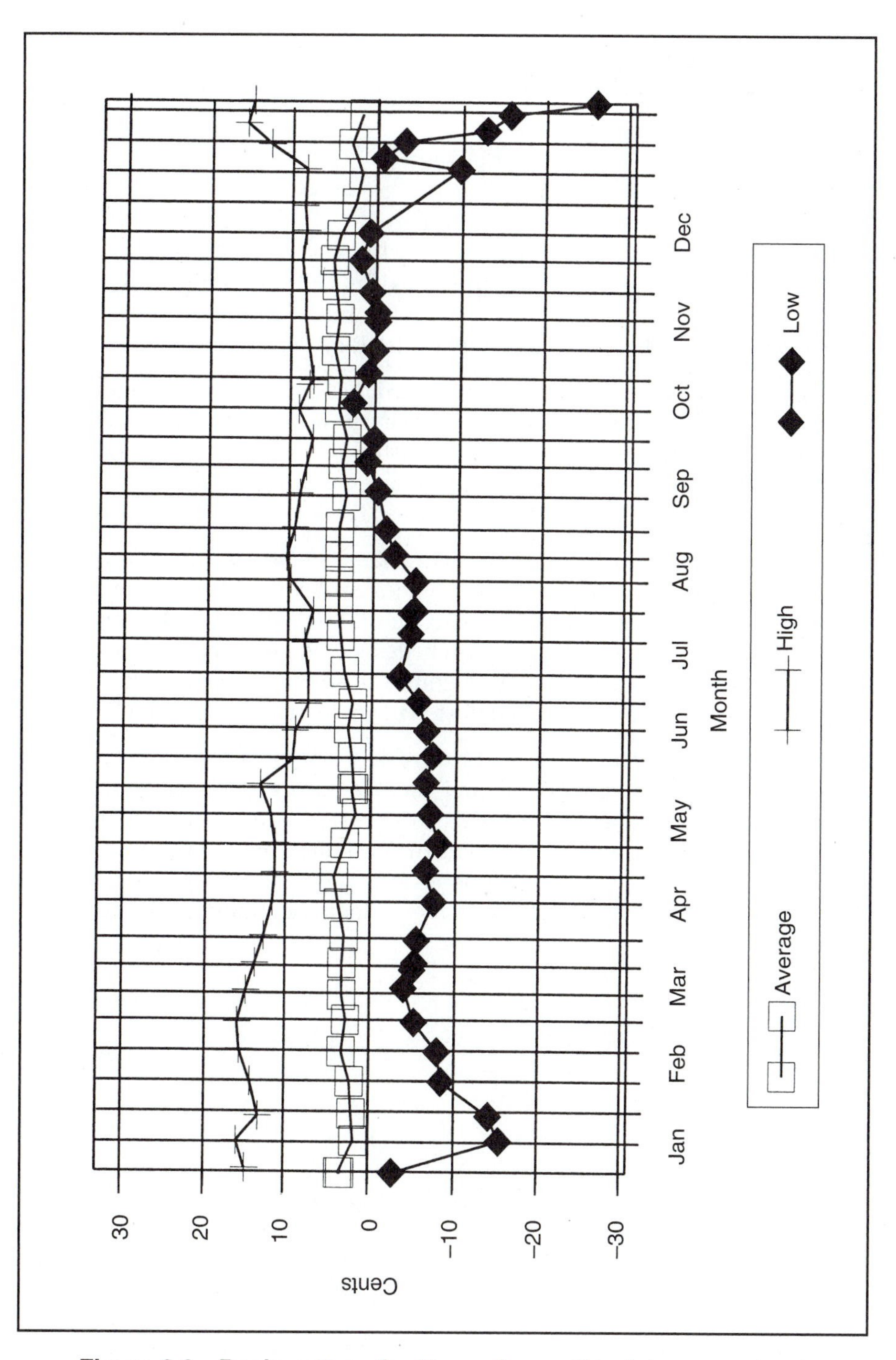

Figure 9.2 Basis pattern for December cotton futures contract (average for years 1973–1982)

cocoa, and orange juice prices are very sensitive to freezes and droughts. Cotton is weather sensitive and has numerous government programs that impact prices.

Problems

1. A crude oil trader has the opportunity to buy cash crude oil for $18.00 per barrel. She could hedge it with the nearby futures at $18.90 per barrel. Just as she is considering the trade, the phone rings and it's a refinery that wants crude oil and is willing to buy it from the trader for "thirty cents off the nearby." Should the trader take both deals? Explain.
2. SouthAm, Inc. is a coffee marketing board that gathers coffee from producers, processes and stores, and then markets it worldwide. Once they sell the coffee internationally, they then return the price they get, less operating costs, to the producers. The producers are pushing the board to consider hedging. How could the board use futures? Explain.
3. Karet, Inc. has several gold mines and a large smelter to processes the ore into gold. They process about 1,000 ounces each week. They accumulate the gold during the week and then sell it on the open world gold market each Saturday. They have a total full-term hedging operation whereby each Saturday they hedge next week's production and simultaneously offset the previous week's hedge. How would the t-account look for Karet, Inc.?
4. You calculate a target price for cotton of $.70 per pound. Your fixed costs of production are $.10 per pound, and variable costs are $.60 per pound. You have already refinanced your fixed costs once with the bank, and they have told you that they will not do so again. You must break even this year or the bank will foreclose. Your brother tells you not to hedge, because you won't make any profit, and besides the outlook for cotton is bullish. Should you hedge?

CHAPTER 10

Hedging Strategies: The Financials

"If I weren't president, I'd be buying stocks now." A businessman replied to me, "Yes, and if you weren't president, I'd be buying them, too."

— John F. Kennedy

Financial futures contracts were first offered in the early 1970s and overall have been extremely successful. In fact the volume of trades has exceeded even the most optimistic estimates. Some have become so successful such as the Treasury Bond contract at the Chicago Board of Trade that additional evening and early morning trading hours have been initiated. As with most futures contracts, speculators account for most of the trading volume, but hedging activity using the financial instruments is growing as more and more businesses recognize the role futures can play in price risk management.

The intent of this chapter is to discuss possible hedging strategies for interest rates, currency, and stock index futures contracts. Each of these topics warrants an entire book to cover all the details and peculiarities adequately (many such advanced books exist, and a selected list is in the recommended readings section). Thus, only the major hedging concepts are discussed in this chapter.

Financial Concepts

There are a two major concepts that are necessary for a good understanding of interest rate futures contracts and hedging: price/rate relationship and yield curves. The following discussion is by no means exhaustive in scope, but merely provides some fundamental information. The selected reading section provides references that can provide more detailed and exhaustive information.

Price/Rate Relationship

A fixed rate bond's price varies inversely with interest rates. If rates increase, the price of the bond decreases, and conversely, if rates decrease, the price of the bond increases. If a two year $1,000 bond (face value) has a coupon of 10%, then the holder will receive $100 annually* for two years and receive the $1,000 principal at the end of two years. If the current market rate (discount rate) is also 10%, then the bond's price is $1,000, as calculated using the discount formula:

$$DV = \sum_{n}^{1} \frac{R}{(1+r)^n}$$

where

DV = Discounted Value of the price of the bond

R = Income per period

r = discount rate

n = time period

* Actually, most government and corporate bonds pay interest semiannually. For this 10% example, the typical bond would pay $50 twice each year.

Therefore,

$$DV = \$1{,}000 = \frac{\$100}{(1.10)} + \frac{\$1{,}100}{(1.10)} = \$90.91 + \$909.09$$

If market rates increase to 12%, then the discount likewise increases to 12% and the bond price falls to $966.20, as:

$$DV = \$966.20 = \frac{\$100}{(1.12)} + \frac{\$1{,}100}{(1.12)} = \$89.29 + \$876.91$$

Viewed another way, new issue bonds would have to be issued with a coupon near the new market levels to be competitive. Thus, the new bond would carry a 12% coupon and return $120 for two years with the principal paid at the end of two years. If the holder of the old 10% bond wanted to sell it for $1,000, he would probably not find any takers because $1,000 would buy the new higher yielding bond issued at 12%. The rational price for the old 10% bond is $966.20, which would yield 12% because of the discounted price. There are many other bond pricing rules, and the reader is referred to the selected readings section for more information.

Yield Curves

A yield curve is simply a graphical representation of the yields of different maturities of the same or near same debt instruments. The same or near same debt instruments are used so that the risk of default does not show up in the yield curve. Thus, a certificate of deposit and a Treasury bill should not be on the same yield curve because the risk of default differential will distort the yield curve. Yield curves should contain the yields of various maturities for certificates of deposit. Likewise, a yield curve could contain Treasury bills, notes, and bonds because they represent the same default risk (none) and are simply different maturities and/or other debt instruments grouped by risk of default.

Yield curves like the one in Figure 10.1 are called normal because they have positive carry, that is, longer maturities have higher yields than short-term maturities. When the curve has negative carry, it is said to be inverted. The cost of carry is the net cost (the stated coupon rate less the value of short-term interest rates). When the coupon rate exceeds the short-term interest rate, a positive carry curve results, and when the interest rate exceeds the coupon rate, a negative carry curve results.

Most yield curves exhibit positive carry. Attitudes towards risk generally favor a positive carry. Holding longer-term instruments is more risky and commands a higher rate, as shown in Figure 10.1. However, expectations about future events and rates have a major influence on yield curves. If expectations are strong that rates will decrease, then investors will hold short-term instruments and the yield curve will express negative carry.

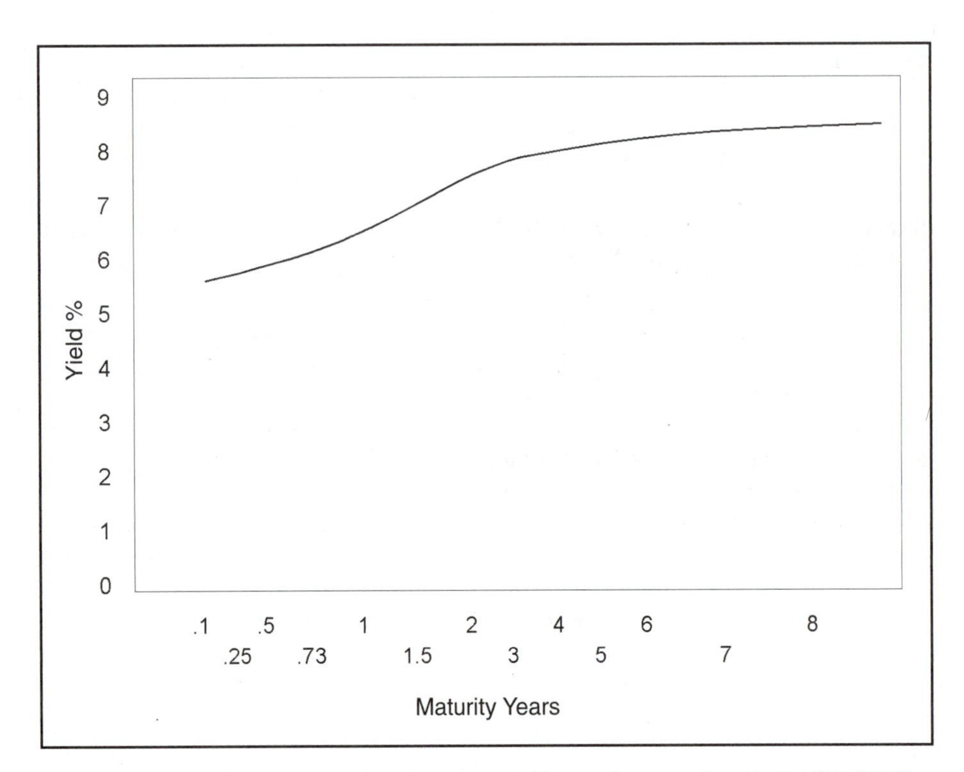

Figure 10.1 Yield curve for Treasury bills and notes for June 25, 1987

Interest Rate Hedging

The first financial futures contracts involving interest rates began trading in October 1975. The timing of the initial futures contract trading corresponded to increased inflation in the U.S. economy and to increased volatility in interest rates and set the stage for additional interest rate futures instruments when interest rate volatility increased dramatically in the late 1970s and early 1980s. The record high levels that interest rates reached, coupled with the increased volatility, promoted a considerable interest among speculators to trade interest rate futures. It also caused financial institutions to begin to look at ways to use these new instruments to hedge and protect against price/rate changes.

Financial institutions for years did not need price protection because of regulations and legislation that controlled rates and limited the kinds of financial instruments that could be offered. An example is the old Federal Reserve Regulation Q, which set a maximum allowable interest rate that could be paid on savings accounts at banks and savings and loan associations. Competition among the various kinds of financial institutions was limited primarily to non-price competition. In fact, many financial executives were called "Dr. Peppers—10-2-4" because they supposedly came to work at 10 a.m., quit at 2 p.m., and were on the golf course by 4 p.m. This is, of course, an unfair appraisal for most of the executives. However, it does point out the lack of attention that was paid to running the financial institutions. The heavily regulated environment promoted a non-price competitive atmosphere, rather than one with price competition. Inflation and deregulation of the late 1970s and early 1980s forced financial institutions to be price competitive. As a result, price changes altered margins and profits. Furthermore, volatile rates that financial institutions were forced to pay for their loanable funds (both to depositors and to the larger regional, national, and/or international wholesale and investment banks) caused a great concern at banks over the cost of funds and thus profit margins. Due to the entirely new, deregulated financial environment of the 1980s, it became critical to try to control the risk of price changes.

The Mechanism of an Interest Rate Hedge

Financial institutions first began looking at hedging as a means to control the risks of price change in financial instruments such as the bonds held as investment and of interest rate changes on both investment bonds and on funds the institution uses to loan to its borrowers. The price/return issue with financial instruments somewhat complicates the hedging process. Most financial instruments have both a return and price component that is directly related to the liquidity of the instrument and its riskiness. Therefore, matching the cash position and the futures contract for a financial instrument is more difficult than for a corn producer or a gold bullion dealer, but it is still possible and this matching, or hedging, process can provide price risk protection.

If a financial institution desires to control interest rate risk with futures contracts, it will soon discover that there actually are no direct interest rate contracts traded anywhere. Instead, the hedger must use Treasury bills, notes, or bonds (depending upon the maturity of its risky asset) or some other financial (bond) instrument and trade bond price rather than interest rate because all Treasury instruments are quoted on a discount basis (thus a price) not on their effective interest rate.

To offset interest rate risk, the financial institution will hedge with an appropriate amount of T-bills, notes, or bonds such that the gain or loss in price offsets interest rate changes. Assume that a bank issues a one-year certificate of deposit to an individual who placed $1,000,000 in the bank with the agreement that the bank will pay interest at the rate of 6 percent per year. The bank will then loan the $1,000,000 to a borrower at an interest rate sufficient to cover the six percent to be paid to the CD owner (called the cost of loanable funds) and any bank operating costs. Suppose, however, that the borrower, anticipating a general interest rate level decline demands a variable rate loan that will be set for six months (at 9 percent) and then be adjusted at the end of that six-month period for the next six months. The bank has the cost of loanable funds fixed at 6 percent for one year, but the revenue from the loan is fixed for only the first six months.

The bank needs to hedge to protect against the risk that the revenue to the bank during the last six months of the loan will decrease, which will occur if interest rates fall. To hedge, the bank would sell T-bill futures, as shown in Table 10.1.

Table 10.1 Bank Hedging to Offset Interest Rate Risk

Date	Cash Position	Futures Position
January 1	$1,000,000 CD at 6% $1,000,000 loan extended at 9%, semiannually adjustable, interest rate Net margin of 3%	Buy one January (next year delivery) T-bill futures at $.94
July 1	Reprice loan at 8% for next six months Net margin of 2% Net loss revenue of $.01	Sell one January T-bill futures at $.95 + $.01

The net result is that although the T-bills are traded in price, the bank protected its interest rate risk with a variable rate loan.

Loan Portfolio Hedging

Banks and other financial institutions try to match the maturities of their loans with the maturities associated with the source of loan funds and maintain the maximum spreads that competition will allow. Matching of maturities is always less than perfect. There are two major types of price risks that loan portfolios have when the maturities do not exactly match. First, the cost of funds for the loan is subject to change during the fixed period of the loan (called the Returns Fixed/Costs Variable Problem). And secondly, the revenue from the loan is subject to change during the fixed period of the source of funds for the loan (called the Costs Fixed/Return Variable Problem). Loan portfolios are often hedged with a futures market instrument that is liquid, reliable, and of roughly the same maturity as the loan portfolio so that the interest rates on both the instrument underlying the futures contract and the loans themselves will respond to the same market forces.

Returns Fixed/Costs Variable Problem

Bank ZZZ has a $1,000,000 portfolio of three-year loans set with an interest rate of 12 percent. Although the best maturity matching process would indicate that the bank should finance the loans with a three-year instrument of some sort, three-year instruments are either hard to acquire or are not in favor with local investors. Consequently, the bank decided to fund the $1,000,000 for the three-year period with one year Certificates of Deposit (which are currently paying an 8 percent rate). The bank has a spread of 4 percent. A deterioration in the spread because of increased costs of the CDs is a very real possibility if market rates increase. The spread is fixed for one year at an adequate 4 percent but is subject to deterioration for the remaining two years of the loans. To combat this potential for deterioration, the bank needs protection against the possibility of rate increases (i.e., T-bill price decreases). If, for example, in one year when the CDs mature the market rates have risen to 9 percent, the bank's spread for the middle third of the life of the loans is down to 3 percent. A proper hedge would mitigate that problem. Conversely, if rates had fallen to 7 percent, the bank's spread would have risen to 5 percent. However, if properly hedged, the bank's spread would have remained at 4 percent. This example is illustrated in Table 10.2.

Table 10.2 Returns Fixed/Costs Variable Hedging Example

Cash Position	Futures Position
First Year — January 1	
Have 3-year loan fixed at 12%	Sell one January T-bill futures* at $.92
Currently funding loan with 1 year CDs at 8% (spread at 4%)	
Second Year — January 1	
Renew CDs for 1 year at 9% (spread at 3%)	Buy one January T-bill futures at $.91 + $.01
	Sell one January T-bill futures at $.91 (at beginning of second year)
Third Year — January 1	
Renew CDs for 1 year at 9% (spread at 3%)	Buy one January T-bill futures at (at beginning of third year) $.90 + $.01
Net Hedge Spread = [4 + (3 + 1) + (3 + 1)] ÷ 3 = 4 percent	

* Treasury bills are sold at a discount. Thus, a $1,000,000 face value 90-day T-bill is sold (in a competitive bidding process) for say $920,000. The U.S. Treasury receives $920,000 and in 90 days pays the holder of the T-bill $1,000,000. The holder has earned $80,000 for 90 days on a $920,000 investment.

Costs Fixed/Returns Variable Problem

A bank also faces the problem that it may have funding sources fixed at a certain maturity and rate and want to loan money at a variable rate. Consider a bank that has $1,000,000 in CDs at 8% with a maturity of one year but has a one-year variable rate loan to a construction firm that is subject to rate change semiannually (fixed for the first six months at 13 percent). The bank's current 5 percent spread for six months is subject to being squeezed if the market rates that drive the loan rate decrease for the second half of the year. The bank, therefore, is worried that rates will decrease (i.e., prices will increase). They can hedge to protect their 5 percent spread. Typically, variable rate loans are tied to some index or major market instrument such as Treasury bills. Table 10.3 shows the Costs Fixed/Returns Variable hedge.

Table 10.3 Cost Fixed/Returns Variable Hedging Example

Cash Position	Futures Position
First Half of Year	
January 1	
Have CD fixed at 8% (1 year maturity)	Buy one June T-bill futures at $.91
Have 1-year loan with variable rate adjusted every 6 months; currently fixed for 1st half at 13% (spread at 5%)	
July 1	
Adjust loan rate down to 12½% (spread at 4½%)	Sell one June T-bill futures at $.915
	+ $.005
Net Hedge Spread = [5 + (4.5 + .5)] ÷ 2 = 5 percent	

Individual Loan Hedging

Individuals and firms also have interest rate risks. If they have borrowed money and entered a variable rate loan contract, the risk is that rates will increase. On the other hand, they may need to borrow money in the future; they thus have a future loan need and must face the risk that rates will be higher in the future. Both these risks can be hedged.

Variable Rate Loan Hedge for Borrowed Funds

If a firm has entered into a two-year operating loan that has a variable rate adjusted every six months, they have one and one-half years of interest rate risk. To protect themselves from this interest rate risk, they can open a hedge that is very similar to hedges for financial institutions. It involves placing a hedge and rolling the hedge every six months when the loan is repriced. This type of rolling hedge is also very adaptable to a selective hedging strategy. Every six months, a decision is made concerning what direction rates are expected to move over the next six months. The hedge is placed if the forecast is for increasing rates, otherwise the position is left unhedged. Table 10.4 shows a rolling hedge for a variable rate loan. The loan rate was adjusted with an index of several values the bank put together, whereas the hedge was with only Treasury bill futures (a cross hedge).

Table 10.4 Variable Rate Rolling Hedge

Cash Position	Futures Position
First 6 months	
Have 2-year loan, adjusted every 6 months; first 6-month period fixed at 13%	Sell January T-bill contracts at $.92 Buy January at $.91 + $.01
Second 6 months	
Loan repriced at 14%	Sell May T-bill at $.94 Buy May at $.935 + $.005
Third 6 months	
Loan repriced at 15%	Sell December T-bill contracts at $.95 Buy at $.94 + $.01
Fourth 6 months	
Loan repriced at 15.5%	

Net Hedge Loan Rate = $[13 + 14(-1) + 15(-\tfrac{1}{2}) + 15.5(-1)] \div 4 = 13.75$ percent

Anticipatory Hedge

If the loan has not already been entered into, but yet the borrower knows that the need for the loan is definite and that she has prequalified for the loan, the borrower faces the risk that rates will increase. To avoid that risk, the borrower can hedge using Treasury bills. For example, assume that an individual plans on starting a new business venture in six months and is currently lining up financing. She finds fixed rate financing for the amount she needs ($1,000,000), but the financial institution is unwilling to commit to a rate now for funds to be lent six months from now. She thinks there is a strong possibility of a rate increase between now and when she commits to the loan, thus she faces a risk that the loan will cost more than she is now budgeting. Proper hedging can offer her protection in the event that rates increase. Table 10.5 illustrates this hedge.

Table 10.5 Anticipatory Hedge

Cash Position	Futures Position
Finds financing in 6 months at 12% (not guaranteed) ($1,000,000)	Sell one December T-bill contract at $.91
6 months later obtains financing at 13%	Buy one December T-bill contract at $.895 + ($.015)
Net Hedge Loan Rate = 13% – 1½% = 11½ percent	

Bond Hedging

The first financial futures contract offered was on the Government National Mortgage Association (GNMAs) certificate in 1975. Treasury bond futures were later offered (in 1977) and quickly exceeded GNMAs and all other financial and nonfinancial futures contracts to become the most popular futures contract of all time. Both these futures contracts offer hedging opportunities, but because of the size of the contract and the sophistication required to properly be in this market, interest rate futures are used mainly by dealers, mortgage bankers, portfolio managers, and construction/development firms. The long and short hedge will be illustrated using a sampling of various hedging situations for the T-bond futures only, because the GNMA futures contract became inactive and stopped trading in the early 1990s.

Long Hedging

Long hedging examples using long-term interest rate futures fall into two general categories that can be exemplified by a dealer who has forward sold bonds and a portfolio manager who must add to a bond portfolio at some future date. First, consider a bond dealer who has an opportunity to sell Treasury bonds today for delivery in two weeks. The bond dealer does not have the bonds and strongly believes prices will move higher in the next few days due to an as yet unannounced but anticipated action by the Federal Reserve Bank. However, the dealer does not want to miss the sale because it is large ($4,000,000) and it will add a new client to the dealer's list. Thus, if handled properly, it represents substantial prestige as well as current and future commissions to the dealer. To structure the transaction, the dealer decides to hedge. First, he quotes the bonds at the current price of $98-16 for delivery in two weeks. He then buys T-bond futures (40 contracts at $100,000 each to equal $4,000,000 commitment) as a hedge against the potential increase in price. The dealer begins the process of trying to buy the bonds for less than $98-16. As the dealer acquires the bonds, he can scale down his hedge by lifting (selling) the amount that has been acquired. Of course, if the dealer had acquired the full $4,000,000 all in one transaction, scaling down the hedge would not have been necessary. The hedge is illustrated in Table 10.6.

Table 10.6 Bond Dealer Long Hedge with a Scale Down

Cash Position	Futures Position
Forward sells bonds at $98-16 for delivery in two weeks.	Buys 40 December T-bond contracts at $98-10
	Scales down ½ of hedge
Buys $2,000,000 in bonds at $99-16	Sells 20 December T-bond contracts at $99-10 + $1-00
Buys $2,000,000 in bonds at $100-16	Sells 20 December T-bond contracts at $100-10 + $2-00
Net Hedge Price = $99-16 – ($1) = $98-16 + $100-16 – ($2) = $98-16 = $197-00/2 = $98-16	

Portfolio managers often face the situation whereby they anticipate or are warned by executives that within a certain period funds will be available for investment. Consider a manager of a university portfolio who has been told that in two months the proceeds from a major fund drive will be available for investment in Treasury bonds. The fund drive is estimated to bring in $2,000,000. The manager firmly believes that in two months Treasury bonds will be trading higher than they are now because he is bearish on interest rates. To take advantage of current bond prices even though she does not have the investable funds, she can hedge long with T-bond futures and protect against the potential increases in price and ensure a higher yield. This is illustrated in Table 10.7. Notice that the basis deterioration in the examples illustrates that the cash loss was not totally offset with a gain in the futures.

Table 10.7 Portfolio Hedge

Cash Position	Futures Position	Basis
September 1 Knows in two months will have $2,000,000 to invest in bonds (Current quote is $98-14 for 20-year 8¼% bonds)	Buys 20 December T-bond contracts at $98-10	$0-04
December 1 Buys $2,000,000 of 20-year 8¼% T-bonds at $112-13	Sells 20 December T-bonds contracts at $110-07 + $11-29	$2-06 change of $2-02
Net Hedge Buying Price = $112-13 – $11-29 = $100-16		

Short Hedging

Like short hedgers in physical commodities, short hedgers in bonds generally own the bonds (hold them in inventory) with the intent of receiving profit upon sale. But because markets do not always do what a bond owner wants it to, a short hedge is employed to eliminate price risk.

A bond dealer typically holds long cash positions and tries to dispose of them at some point in the future at a profit. Holding the long cash position (an inventory) exposes the dealer to the risk of decreasing cash prices. A short hedge can offer protection against the decreasing inventory value. A hedge that protects against price decrease (a decrease in inventory value) is illustrated in Table 10.8 with a basis deterioration.

Table 10.8 Bond Dealer Inventory Hedge

Cash Position	Futures Position	Basis
May 10		
Holds $1,000,000 20-year T-bonds priced at $94-26	Sell 10 T-bond futures at $90-26	$4-00
May 30		
Sell when prices for bonds fall to	Buy 10 T-bond futures at	
$90	$86-26	$3-06
– $4-26	+ $4-00	change of $0-26
Net Hedge Selling Price = $90 + $4 = $94-00		

Net Hedge

A financial institution often owns bonds as a means to turn deposits or other cash reserves into an interest-earning asset. Consider an institution that has a block of T-bonds that it wants to protect from a potential decrease in value due to declines in prices. A simple inventory hedge will protect against the decline. Table 10.9 shows the inventory hedge with both increases and decreases in prices. Notice that the financial institution is protected no matter what happens to prices. Obviously, the financial institution would rather not be hedged when prices increased so that they could take advantage of the increased value of the block of T-bonds. Unless the financial institution has a good selective hedging system, they must be content when they hedge with the outcome or not hedge and absorb the risk of decreasing inventory values. The hedge prevented the inventory value from decreasing and maintained the inventory value when prices increased. Only opportunity profits are lost when prices increase; although they may appear to be real losses to the investor, the loss of these profits should be viewed as the insurance premium for avoiding losses that might lead to the failure of the business.

Table 10.9 Financial Institution Inventory Hedge

Cash Position	Futures Position
January 15	
Has $5,000,000 T-bond block to protect at $89-00	Sells 50 June T-bond futures contracts at $78-04
Price Decrease	
March 24	
Price at $78-22	Buys 50 June T-bond contracts at $69.00
	+ $9-04
Block of T-bond has declined in value $515,625.00	Gain in value of $456,250.00
Price Increase	
March 24	
Price at $93-00	Buys 50 June T-bond contracts at $82-00
	– $3-28
Block of T-bond has increased in value $200,000	Loss in value of $193,750

Spread Hedging

Spreading is most often used by speculators. However, hedgers are beginning more and more to use the same concepts to protect cash relationships. The financial futures offer many spread hedging opportunities to the astute hedger. Spread hedging can involve spreading the relationship between delivery months of the same commodity (January T-bonds and March T-bonds), or between different but similar commodities at different locations (January T-bonds [CBT] and January T-bills [IMM]).

Spread hedging the financial complex is conceptually the same as spreading as a speculator except that hedgers are trying to protect the change in the relationship of a cash position. Consider an investor who does not have a portfolio but does have investments in T-bonds and T-bills. He made the investment because of a relationship between the long-term rate and the short-term rate. Should those relationships get out of line with each other, the investor must reconsider the investment and make decisions about liquidating one and adding to the other or simply changing the mix by investing in other instruments. Rather than change the mix or investment, at least in the near term (a few months), the investor can spread hedge to protect the relationship.

Consider an individual investor who has invested in T-bonds and T-bills based upon the relationship that the difference between the two has been approximately $4 for the last several months. If the relationship changes while she has the investment, then our investor can put on a spread hedge to protect the relationship. If in a few weeks the spread widens to $5, she can put on a spread hedge by buying T-bonds and selling T-bills. When the spread goes back to the normal $4, the spread hedge is lifted. She has earned an extra $1 without changing the cash position mix. If the spread did not return to normal in a reasonable time, she could liquidate the spread hedge and not be out anything other than commissions and interest on margins. Alternatively, if the spread falls below $4, a **reverse spread** hedge could be placed by selling T-bonds and buying T-bills. Tables 10.10 and 10.11 show both a normal spread hedge and a reverse spread hedge.

Table 10.10 Spread Hedge Using T-Bonds and T-Bills

Cash Position	Futures Position
Has $100,000 face value T-bonds at $94-00	
Has $100,000 face value T-bills at $.98	
$4/$100 face value normal spread Widens to $5	
	Buy T-bonds at $93-16 ($93.50) Sell T-bills at $.985 ($98.50)
	Sell T-bonds at $94 Buy T-bills at $.98
	T-bond net + (16/32) T-bill net + $.50 Net $1

Table 10.11 Reverse Spread Hedge Using T-Bonds and T-Bills

Cash Position	Futures Position
Has $100,000 face value T-bonds at $94-00	
Has $100,000 face value T-bills at $.98	
$4/$100 face value normal spread Spread goes to $3	
	Sell T-bond futures at $94-16 ($94.50) Buy T-bills futures at $.975 ($97.50)
	Buy T-bond futures at $94 ($94.00) Sell T-bill futures at $.98 ($98.00)
	T-bond net + 16/32 T-bill net + $.50 Net $1.00

Equalizing the Hedge

A major problem with hedging interest rate futures is equalizing the quantity of futures contracts with the cash position such that gains and losses are roughly equal. Adjusting the quantity of futures contracts to the cash position to gain dollar equivalence in price movements requires a hedging ratio. The ratio is calculated based upon the level of the futures position relative to the level of the cash position. Equalizing the hedge is also known as weighting the hedge.

Equalizing the hedge is essentially trying to stabilize basis movements. This is particularly important with financial instruments because of maturity and risk differences. It is very rare when the cash instruments match the futures contract terms exactly. Thus, almost all interest rate hedging is truly cross hedging. Cross hedging is discussed more fully in Chapter 11; however, a brief introduction is provided in this chapter for illustrative purposes.

The idea behind equalizing the hedge position is that if the cash instrument is a 180-day certificate of deposit that is going to be rolled in 180 days and the hedge is placed with Treasury bills, then risk and maturity differentials exist. Treasury bills are considered riskless, but even the best CDs have some element of risk of default. Also, the CD has a maturity of 180 days, and the T-bills are 91-day maturity. Different maturities usually have different interest rate levels and, most importantly, different rates of change in interest rate levels. That is, you would expect most of the time that a 180-day rate would be expected to change in price (and thus yield) at a rate that is different than for a 91-day rate. Because of the mismatch in maturities and risk levels of the instruments, basis movements can cause the hedge to be very ineffective and in some cases disastrous.

Effective equalization of hedges depends upon solid knowledge of both the cash instrument and the futures contract and knowledge of how the prices have moved in the past. There is, unfortunately, no sure method to calculate the hedging ratio. The optimal hedging ratio changes as the separate markets change. In fact, many hedgers ignore the equalizing problem because it is so complex and instead choose to merely hedge (or cross hedge) to try to get some protection. The equalizing problem has caused exchanges to offer many different types of futures contracts to help mitigate the problem. But in fact, this may actually cause the problem to worsen. Many cash traders don't hedge because of the equalizing problem, so the exchange offers a new contract designed to help. But because they aren't experienced hedgers, they ignore the new contract and it dies or is very thin.

If some thought goes into the hedging process, equalizing can be effective. For example, if a simple analysis of past price movements shows that when 180-day CD rates fall one point while the March T-bill futures price increases only 1/10 of 1 percent, then the futures position will not fully protect the cash position unless the quantities can be adjusted. For example, consider a 180-day CD cash position that will be rolled and a hedge with T-bills (shown in Table 10.12). The cash position had an opportunity loss of $5,000 because rates fell one full point; however, the futures position gained only $1,000. The hedge was placed such that both the cash and futures were equal in dollar amounts, but not equal in dollar changes. The hedging ratio is 5:1 in this situation—$5,000,000 (five T-bill futures contracts) are needed to get the same dollar change in the cash position. Thus, proper weighting of the hedge is necessary for good hedging performance. This requires some basic financial knowledge and past price information.

Table 10.12 An Equal Hedge with T-Bills

Cash Position	Futures Position
Have $1,000,000 CD at 8%, 180 days	Buy one December T-bill at $.91
Roll CD in 180 days Have $1,000,000 at 7%	Sell one December T-bill at $.911
Opportunity loss of 1 point/6 months or $5,000	+ $.001 ($1,000)
	Need five December T-bill contracts to cover 1 point movement in the CD

Currency Hedging

Futures contracts on international currencies began trading in the late spring of 1972 (May 16). This proved to be excellent timing because the United States had abandoned the gold standard and the Bretton Woods agreement for fixed exchange rates. Thus, the dollar was left to float or sink relative to other currencies, at the will of market forces. The dollar began a roller coaster ride that continues today, despite calls for a full or partial return to a fixed exchange rate system and even a return to the gold standard. The increased volatility of the dollar as measured against major currencies added tremendous currency risk to international transactions. United States, non-U.S., and multinational firms alike have seen profits erased as currency values change. Clearly, some firms have also profited by these changes, but the potential losses due to exchange rate problems are usually viewed more as an impediment to international trade than as a profit opportunity.

The foreign currency market has a very active and liquid forward contract market that allows firms to contract forward for currency purchases or sales. However, there are many differences between the forward market and the futures market, as illustrated in Figure 10.2. The examples in this chapter concentrate on the use of futures contracts to hedge rather than use the forward cash market as a risk management tool.

Short Hedging

A U.S. firm that has an export sale to Great Britain with payment to be made in British Pounds Sterling faces the risk that the pound, relative to the dollar, will depreciate. That is, assume the current rate of $1.40 per 1 pound sterling moves

COMPARISON OF THE CURRENT FUTURES MARKET WITH THE INTERBANK MARKET

Dealers in the Interbank market offer forward contracts, which provide benefits that are similar to those of futures contracts. Although similar in concept, the two types of contracts have distinctions as well. The following comparisons summarize the major differences between the two markets.

Futures Market	Spot & Forward Interbank Market
1. Trading is conducted in a competitive arena by "open outcry" of bids, offers, and amounts.	1. Trading is done by telephone or telex, with banks generally dealing directly with other banks, foreign exchange brokers, or corporations.
2. Participants are either buyers or sellers of a contract at a single, specified price at any given time.	2. Participants usually make two-sided markets (quoting two prices that indicate a willingness to buy at a lower price and sell at a higher price), for both spot and forward prices.
3. Non-member participants deal through brokers (Exchange members), who represent them on the trading floor.	3. Participants deal on a principal-to-principal basis either directly or through brokers.
4. Market participants usually are unknown to one another, except when a firm is trading its own account through its own brokers on the trading floor.	4. Participants in each transaction always know the other trading party.
5. Participants include banks, corporations, financial institutions, individual investors, and speculators.	5. Participants are banks dealing with each other, and other major commercial entities. Access for individuals and smaller firms is limited.
6. Trading prices of currency futures are disseminated continuously by the CME.	6. Indicated bids and offers, as opposed to actual prices, are available throughout the Interbank market.
7. The Exchange's Clearing House becomes the opposite side to each cleared transaction; therefore, demands of monitoring credit risk are substantially reduced.	7. Each counter party with whom a dealer does business must be examined individually as a credit risk, and credit limits set for each. As such, there may be a wide range of credit capabilities of participants.
8. Margins are required of all participants.	8. Margins are not required by banks dealing with other banks, although for smaller non-bank customers, margins may be required on certain occasions.
9. Settlements are made daily via the Exchange's Clearing House. Gains on position values may be withdrawn and losses are collected daily.	9. Settlement takes place two days after the soft transaction (one day in the U.S. for the Canadian Dollar and Mexican Peso). For forward transactions, gains or losses are realized on the settlement date.
10. A small percentage (usually less than 1 percent) of all contracts traded results in actual delivery.	10. The majority of trades result in delivery.
11. All positions, whether long or short, can be liquidated easily.	11. Forward positions are not easily offset or transferred to other participants.
12. Standardized dates are used for all contract months, concentrating liquidity to produce maximum price competition.	12. Settlement of forward contracts can be at any date agreed upon between the buyer and seller.

Figure 10.2 Comparison of forward markets and futures markets for foreign currency *(Source: IMM)*

to \$1.35 per 1 pound sterling. If the U.S. firm has agreed to receive 100,000 pound sterling for the merchandise, then the loss in value is \$5,000—they could have received \$140,000 at the old exchange rate, but with the new rate they only receive \$135,000. (This difference could be critical, especially if the profit margin or the merchandise was less than \$5,000 or if the firm made a pricing decision at \$1.40 to capture a specific profit margin.) They need to protect against the dollar appreciating (the pound depreciating). This can be done with a short hedge, as shown in Table 10.13. Notice that whether the dollar appreciated or depreciated, the net hedged exchange rate remained at \$1.40 and the firm was protected.

Table 10.13 Example of Short Foreign Exchange Hedge with British Pounds

Cash Position	Futures Position
May 1	
U.S. firm agrees to sale in Great Britain.	Sell British pound futures at \$1.40
Sales price 100,000 pounds sterling	
(current exchange rate is \$1.40/pound)	
Dollar Appreciating	
May 15	
Receives 100,000 pounds	Buy British pound futures at
Converts to U.S. dollars at \$1.35/pound	\$1.35
Received \$135,000	+ \$.05
\$5,000 opportunity loss	
Net Hedge Price = \$1.35 + \$.05 = \$1.40	
Dollar Depreciating	
May 15	
Receives 100,000 pounds	Buy British pound futures at
Converts to U.S. dollars at \$1.45/pound	\$1.45
Received \$145,000	– \$.05
\$5,000 opportunity gain	
Net Hedged Price = \$1.45 – \$.05 = \$1.40	

A similar situation exists for individuals and/or firms that send money out of the United States and then bring it back. The risk is that the dollar will appreciate relative to the country's currency for which they have temporarily exchanged. They run the risk, if they exchange their dollars, that when they want to bring the money back to the United States, the currency will not buy the same amount of dollars that was originally exchanged. Consider a situation involving an investor who sees an investment opportunity in Canada. The investor figures that she can earn a higher return for three months by investing in Canada versus the United States. The risk the investor runs is that when the money is converted in three months, the exchange rate will be such that the extra profit earned on the investment in Canada will be lost to exchange rate changes. For example, if $100,000 Canadian is purchased at a rate of $.90 U.S. for $1.00 Canadian and invested in Canada, the risk is that in three months the exchange rate will be at $.88 U.S. for $1.00 Canadian. Thus, when the investor brings the $100,000 Canadian to the exchange market, she will receive only $88,000 U.S.—a $2,000 loss. Furthermore, if the investment was successful, then more than $100,000 Canadian will be exchanged. If the investor earned $3,000 Canadian on the $100,000 Canadian investment, then the investor will receive $90,640, and she paid $90,000 for the $100,000 Canadian. Her investment earned only $640 for three months rather than $2,700 (3,000 × $.90). To protect the projected investment earnings difference, a short hedge could be placed as shown in Table 10.14.

Table 10.14 Example of Short Foreign Exchange Hedge with Canadian Dollars

Cash	Futures
June 1	
Buys Canadian dollars and invests in Canada ($100,000 Canadian) Rate at $.90 U.S./$1 Canadian ($90,000 U.S. cost)	Sell Canadian dollar futures at $.90
September 1	
Sells $100,000 Canadian plus earnings of $3,000 Canadian and buys U.S. dollars at $.88 U.S./$1 Canadian ($90,640 U.S.)	Buys Canadian dollar futures at $.88 + $.02 ($.02 × 100,000 = +$2,000)
+ $640	
	Net Gain = $2,640 U.S. gain

Long Hedging

Long hedging protects against the dollar depreciating relative to other currencies. This typically involves selling a foreign currency and buying dollars and then later selling the dollars and buying the foreign currency. A firm that borrows 100,000 pounds in England and transfers them for the duration of the loan to the United States faces a foreign exchange risk that the dollar will depreciate while the loan is in effect. If he borrows the 100,000 pounds and converts them to dollars at a rate of 1.4, he would have $140,000. When he has to convert back to pounds to pay off the loan, the risk is that the rate will be at 1.5. If it is at 1.5, then the $140,000 would convert to only 93,333 pounds—a 6,667 pound shortfall to pay off the loan principal, not to mention the earned interest. A long proper hedge would protect against that shortfall, as illustrated in Table 10.15.

Table 10.15 Example of Long Foreign Exchange Hedge with British Pounds

Cash	Futures
May 1	
Borrows 100,000 pounds Converts to U.S. dollars at 1.4 rate = $140,000	Buy British pound futures at $1.40
August 1	
Sell U.S. dollars and buys pounds at 1.5 rate	Sell British pound futures at $1.50 + $.10
$140,000/1.5 = 93,333 pounds – 6,667 pounds	($.10 × 100,000 = $10,000 gain)
(1.5 × 6,667 = $10,000 loss)	

A similar problem exists for firms that want to transfer funds temporarily from a subsidiary in West Germany to the United States and then back to Germany after a period of time. The risk is that the dollar will depreciate during the period that the dollars are held. A proper long hedge with German Mark futures would protect against the depreciating dollar (illustrated in Table 10.16).

Table 10.16 Example of Long Foreign Exchange Hedge with German Marks

Cash	Futures
May 1	
Sell marks at $.40 U.S. and buy U.S. dollars	Buy marks futures at $.40 U.S.
August 1	
Sell U.S. dollars and buy marks at $.42 U.S./1 Mark – $.02	Sell marks futures at $.42 U.S. + $.02
Net Hedge Price = $.42 – $.02 = $.40	

The very simple rule for hedging with foreign currency futures traded on U.S. exchanges is this: Short hedge when protection is needed for an appreciating dollar relative to another currency and long hedge when protection is needed for a depreciating dollar relative to another currency.

Basis Aspects

Basis in foreign exchange is dominated by interest rates and expectations about inflation, balance of trade, and purchasing power. Interest rates between countries determine a base difference between a spot exchange rate and a futures contract. Using the assumption of no restrictions of trade between countries, then the forward rates will inversely reflect the difference between the two countries' interest rates. That is to say, if the interest rate in the United States is lower than the rate in Great Britain, then the futures contracts for pounds will be discounted to reflect the higher interest rates in Great Britain. If interest rates in Great Britain are 1 percent higher than in the United States, the British pound will be forward discounted by 1 percent.

Why is this generally the case and what are the forces that cause it? First, rational investors will attempt to take advantage of the higher yields in Great Britain. They convert dollars into pounds, transfer the pounds to take advantage of the higher rate, then transfer the pounds to dollars and return them to the United States. Equilibrium between supply and demand (the theory and the graphs are almost exactly the same for British Pounds as they are for any domestic physical commodity) would indicate that the first exchange of dollars for pounds has a tendency to drive the price of dollars lower relative to the pound, whereas the second action has a tendency to drive the price of dollars higher relative to the pound—thus the forward rate reflects this second action, and the pound's forward rate is trading at a discount to the dollar. Basis trading arbitragers will keep this relationship fairly stable and constant.

The interest rate differential component of basis (often called the interest rate parity concept) can be distorted when the markets start reflecting strong expectations about inflation changes and major changes in the balance of trade between countries. When significant events and economic conditions change, the basis for foreign exchange will temporarily not reflect the inverted interest rate differential and will reflect undetermined expectations.

Currency Concepts

Some of the basic determinants of exchange rates are discussed as well as some of the major dimensions of international trade and economics.

Exchange Rate Determinants

The broad long run determinant of exchange rates is called the theory of exchange parity (also called Purchasing Power Parity). The idea of the theory is that the exchange rate between two countries should be the ratio of the price levels of the two countries. If $10 in the United States buys the same thing as 5 pounds sterling does in Great Britain, then the exchange rate will be $2 to 1 pound sterling. If it goes to 2.5 to 1, then Great Britons will import more from the United States, and if it goes to 1.8, Americans will import more from Great Britain. Of course, there are many factors that prevent the exchange parity theory from operating perfectly.

Transportation and transaction costs will prevent the exchange parity concept from reflecting the price level ratio. Similarly, tariffs and quotas place restrictions on goods and services moving between two countries, and consequently the exchange parity will not reflect the exchange rates as the ratio of price levels. Furthermore, exchange rates tend to reflect the ratio of the prices of goods and services traded internationally rather than the whole basket of goods and services in each country.

Short-term determinants of the exchange rate between two countries are interest rates, inflation, and economic activity.

Interest Rates

When one country's interest rate rises relative to another's, then funds flow to the country with the higher rate. This causes an increase in the demand for the currency of the higher rate country and moves the capital account towards a surplus, which in turn sets in motion an appreciation of the higher rate country's currency, vice versa. Conversely, a decrease in the demand for the currency of a country, caused by a decline in interest rates relative to the rest of the world, moves the capital account toward a deficit and generates a depreciation of the lower rate country's currency.

Inflation Rates

Expectations concerning inflation have a strong impact on exchange rates. The discussion of interest rates earlier in this chapter was based on real rates. If a country's interest rate increases because of inflation, this may not be attractive to international investors depending upon the riskiness of the assets. Other things being equal, higher rates of inflation relative to another country will have a tendency to depreciate that country's currency.

Economic Activity

Two countries will not typically experience the same rate of growth or concentration of their economic activity. Because the two countries' growth rates are not moving simultaneously, their exchange rates will fluctuate. The two countries' capital and current accounts will be moving towards or away from surplus and thus changes in exchange rates. One scenario says that as a country experiences growth, the current account will move towards a deficit because in the short run the country will be importing more than they export, whereas an economic slump or contraction will move a country's current account toward a surplus as the country exports more than it imports. Thus, the country that has economic growth will see its currency depreciate, and a slumping economy will suggest currency appreciation.

Index Futures

The success of the currency and interest rate futures naturally led to other futures instruments that could be used to offer investors profit potential and hedging. The stock market was an obvious choice. Millions of investors use the stock market for risky investments as well as structured portfolios. Additionally several other indexes have been or are being considered for possible futures contracts such as economic growth (major market indicators), housing starts, and automobile sales. A Consumer Price Index futures contract existed and traded briefly.

The concept of an index rather than a single commodity poses two major problems for hedgers. First, it is difficult to hedge a single commodity with an index because the cash position doesn't match the futures position. Secondly, it is difficult to structure a basket of commodities that exactly matches the futures. Basis movements are the major problem of index hedging. To use index futures properly in a hedging program requires empirical research to decide which index futures is the most appropriate either for a single commodity hedge or for a portfolio of commodities.

Stock Index Hedging

The price of an individual stock can change due to factors unique to the company (profits, growth, or takeovers) or factors outside of the company (money supply, balance of payments, or taxes). The first type of risk is called unsystematic risk; the risk of price change due to outside factors is called systematic or market risk.

Investors who want to hold several stocks can control unsystematic risk by grouping stocks into a portfolio. This portfolio should have enough different types of stocks so that when one stock experienced a decline in price, another would have a gain. Thus a properly constructed portfolio of stocks would eliminate unsystematic risk. The portfolio would then only be subject to factors outside the control of the individual investor—systematic or market risk.

Systematic risk can be controlled by hedging with stock index futures. Several stock index futures contracts exist that can help portfolio managers, stock traders, and stock owners handle the risk of markets changing. Stock owners face the risk that their long cash position will decline in value, and potential stock purchasers are worried that stock prices will increase.

Short Hedging

Consider a firm that has a portfolio currently valued at $1,000,000. They believe that the stock market is gong to turn bearish and their portfolio will decline in value. If a major decline occurs, then the portfolio may have to be restructured, but regardless of the restructuring, they will suffer a loss. They can protect the cash portfolio by selling stock index futures. Likewise, if their bearish fears prove wrong and the market increases, then they give up any increase (as illustrated in Table 10.17).

Table 10.17 Short Hedging a Stock Portfolio

Cash Position	Futures Position
July 1	
Portfolio value at $1 million	Sell four December S&P 500 stock index futures at $250.00
Decline	
November 15	
General stock market has declined in value. Portfolio now at	Buy four December S&P 500 stock index futures at
$975,000	$238.00
– $25,000	+ $12 × 500 = $6,000/contract
	Total gain = $24,000
Increase	
November 15	
General stock market has increased in value. Portfolio now at	Buy four December S&P 500 stock index futures at
$1,025,000	$252.00
+ $25,000	– $12 × 500 = $6,000/contract
	Total loss = $24,000

Another short stock index hedge is used by stock traders who hold large amounts of stocks (sometimes called market makers). They have a similar problem to the portfolio manager in that they risk a decrease in the value of their cash inventory. Often this type of hedging carries more basis risk than a well-structured portfolio hedge because the makeup of the trader's inventory is less structured and more random than the typical portfolio. The trader may have ten different stocks, but do they follow an index with any consistency? Non-portfolio hedging with stock index futures imparts a considerable basis risk that may or may not be less than the absolute price movements. As an example, consider a trader who has three industrial stocks and two utilities in inventory (purchased for $500,000). This cash position is hedged with four S&P stock index futures contracts, as illustrated in Table 10.18. The example shows both a basis movement for and against the hedger. Basis movements of this magnitude are not uncommon in non-portfolio hedging situations.

Table 10.18 Non-Portfolio Short Hedge Example

Cash Position	Futures Position	Basis
May 1		
Trader has five stocks inventoried at a value of $500,000	Sell four June S&P 500 stock index futures at $260 (total value = 500 × $260 × 4 = $520,000)	$20,000
	Basis Improvement	
May 15		
Inventory has declined in value to $480,000	Buy four June S&P 500 stock index futures at $248 (total value = 500 × $248 × 4 = $496,000)	$16,000 change of $4,000
	Net Inventory Value = $480,000 + $24,000 = $504,000	
	Basis Deterioration	
May 15		
Inventory has declined in value to $480,000	Buy four June S&P 500 stock index futures at $252 (total value = 500 × $252 × 4 = $504,000)	$24,000 change of $4,000
	Net Inventory Value = $480,000 + $16,000 = $496,000	

Long Hedging

Long hedging with stock indexes falls into two major categories: hedging an anticipated purchase of stocks and short selling of stocks. Often portfolio managers and traders know in advance that purchases will be made at some point in the future. They can therefore use futures contracts to protect against price increases. Short sellers are worried that cash stock prices will increase and need protection.

A portfolio manager is told that $500,000 will be available for investment through portfolio expansion in one month. The manager tries to maximize the number of stocks in the portfolio without sacrificing dividends or buying very risky stocks. The planned expansion will be with stocks currently in the portfolio. What the manager wants to do in one month is buy the largest number of stocks available for $500,000 given the current mix of stocks in the portfolio. If stock prices increase, then the number of stocks that can be purchased will decrease. (The portfolio is proxied out—thus the need for maximizing the number of stocks in the portfolio. Each share carries a vote, and the vote can be transferred [a parity], thus the number of votes or proxies has value in corporate voting.) This anticipated cash position can be hedged as illustrated in Table 10.19.

Table 10.19 Anticipated Purchase Long Hedge Example

Cash	Futures
January 1	
Anticipates in one month \$500,000 will be available to purchase several stocks to add to portfolio (could buy the stocks today at an average price of \$68/share or 7,353 shares)	Buy four KC value line indexes at \$260
February 1	
Receives the \$500,000 and finds the average price has risen to \$73/share and therefore can buy only 6,849 shares	Sell four KC value line indexes at \$276 $\$16 \times 500 \times 4 = \$32,000$
	\$32,000/\$73 = 438 shares
Net Shares Purchased = 6,849 + 438 = 7,287	

A short seller of stock borrows the stock from someone such as a broker (or trader). The intent is that if prices fall, the trader can purchase the stock and deliver against the short sale and have a profit. However, if prices increase, then the short seller faces a loss because the stock will have to be purchased at a higher price (often higher than they were sold short). This short sale can be properly protected with a long hedge, as shown in Table 10.20.

Table 10.20 Short Sale of Stock Hedge Example

Cash	Futures	Basis
February 15		
Sell stock short at \$60/share (1,500 shares) Value = \$90,000	Buy one NYSE composite index futures at \$175 Value = \$87,500	\$2,500
March 1		
Stock has increased in price to \$65 Value = \$97,000	Sell NYSE composite index futures at \$186 Value = \$93,000 + \$5,500	\$4,500 change of \$2,000
Loss to seller = \$7,500	Gain = \$5,500	

Basis Aspects

Basis for stock index futures is confusing and somewhat elusive. All stock index futures are cash settled. The opportunity to **arbitrage** under- and overpriced cash and futures is eliminated through cash settle. With gold, for example, if the futures price were greater than the cash price plus carrying charges, then the futures could be sold and the cash purchased and the difference would yield the arbitrager a profit. However, when the arbitrager cannot tender the cash for a futures delivery or acceptance of delivery, then arbitrage opportunities vanish. As a consequence, the spread between the two different delivery months and basis itself can only be stated theoretically because arbitrage is difficult. It is possible, however, to construct a set of stocks that follow the indexes very closely and thus profit by arbitraging the over- and underprice situations. In fact, some traders have constructed such sets of stocks, and these are appropriately called index funds because they track the futures indexes closely. These index funds provide some arbitrage opportunities and tend to provide some market actions that keep the futures indexes and the underlying cash stocks tracking closely together.

The theoretical basis for stock indexes is simply the cost of carry. The cost of carry is modified for stock indexes versus a pure inventory commodity such as gold by the dividend factor. Stocks, in addition to changes in prices, have the potential of accrued dividends. Therefore, the cost to carry or hold stocks is reduced by the amount of dividends. Any forward price for stocks would reflect the financing charge less dividends and basis would be the difference between the two prices, as

$$\begin{aligned}\text{Basis} &= \text{SP} - \text{FP} \\ &= \text{FC} - \text{D}\end{aligned}$$

where

SP = Spot Price at time t

FP = Futures Price for time t + 1 at time t

FC = Financing Costs for period t to t + 1

D = Stock Dividends for period t to t + 1

Arbitragers and rational traders use the theoretical model as a rough guide in making hedging and investing decisions. If, for example, the faraway stock index futures were trading at the same or near same levels as the nearby that was about to expire, then a rational trader would be better off holding a long futures than an actual cash position (assuming a positive cost of carry). These and similar actions will tend to keep the index futures and the cash markets from becoming widely distorted even in the absence of strong arbitrage.

Swaps and Other Derivatives

The swap market developed in the late 1970s and has expanded rapidly since. Likewise other derivative instruments have developed just as quickly. Swaps are discussed separately from other derivatives simply because they are very popular and more standardized than other derivatives.

Swap

Simply put, a swap is an agreement between two parties (called counter parties) to exchange payments with each other. There are four general classifications of swaps: interest rates, currencies, commodities, and equity. These types of swaps are used to handle the risk that values will change from what is expected or needed.

Because it is often difficult for individuals to find the parties who have needs that can be matched with theirs, swap dealers and brokers have emerged. Dealers take the other side of each deal and earn a spread difference for their efforts. Brokers on the other hand simply match up counterparties and receive a fee for their efforts.

For example, the most popular interest rate swap is the fixed for floating where one counterparty exchanges (swaps) her floating interest rate payment to another counterparty for his fixed rate payment. Generally, the underlying principal is not swapped—just the payment schedules.

If one firm has access to relatively cheap fixed rate financing but desires a floating rate because they think rates may change over time in their favor, they would find someone who did not have access to as good a deal on fixed rate financing and swap payment schedules.

Swaps are very flexible. They can be constructed to match long-term needs over several time periods, for various amounts, and can be tailor made to suit individual company needs. If properly done, they can control interest rate risks, currency risks, equity value risks, and commodity price risks. They should not be viewed like futures, however. Futures are more standardized and less flexible. Therein lies one of the major reasons for swap popularity—they can be made into just about any instrument that the human mind can construct. However, it must be understood that futures are openly traded and backed by the clearing corporation at each exchange. Default risk for futures is almost nil. Not so with swaps. They are only as good as the parties and dealers who construct and trade them.

Other Derivatives

It seems like everything going on in the 1990s concerning financing involves the word "derivative." Reports that companies are handling their risks with derivatives are always in the news, and just as often in the news is word that someone lost a bundle using derivatives.

A derivative is a security whose value or returns is derived from some other underlying asset or security. Obviously futures are derivatives—they are derived from the underlying commodity or security. An option on futures is a double derivative. Its value is derived from a futures whose value is derived from the underlying commodity or security.

In addition to swaps, other derivatives involve simple combinations of futures and options, but some become very exotic and complicated. The reader is referred to the reading list for books that cover the more-complicated derivatives. As with swaps, certain derivatives can be structured to be very individual for businesses and certain types of risks. It must be pointed out that the complicated derivatives have two major risks: the risk of default, and the risk that they do not cover the risk needing protection.

The first risk is the same with swaps—they are only as good as the parties and dealers involved. The second risk is that they may not protect the risk they were intended to and in fact may not be a price risk management tool at all. Many complicated derivatives are simply complicated, speculative strategies. Hedgers should enter this water carefully—and fully informed.

PROBLEMS

1. Big Bank has loaned a customer $1,000,000 for two years at a fixed rate of 11%. The bank has financed the loan with another customer's Certificate of Deposit (CD). The CD matures in six months and has a current rate of 5%. The bank has established a gross margin of 6 percent for the next six months. Set up a t-account showing how the bank could hedge its risks on this situation.

2. International Investors has firms both in Canada and the United States. They convert $1,000,000 Canadian to U.S. dollars at an exchange rate of 0.72 ($1 Canadian buys $.72 U.S.). They send the $720,000 U.S. to the United States as a short-term loan to the U.S. office. The U.S. office has agreed to return the $1,000,000 Canadian to the Canadian office plus 10% interest. Set up the t-account hedge and explain how the U.S. office can protect itself.

3. Carlos works for the development office of a university. The university is currently having a fund raising drive to raise an additional $3,000,000 to put into the foundation's endowment. The endowment is currently a very diversified portfolio of various stocks. The president of the university has told Carlos that in two months the $3,000,000 will be available and it should be put into diversified stocks similar to the current portfolio. Does Carlos face any price risks? Could he hedge? If so, how?

4. The current interest rate on short-term government Treasury bills in Canada is 6%, but the new issue of similar Treasury bills in the United States carries a rate of only 5%. The current exchange rate is 0.74 ($1 Canadian buys $.74 U.S.). Both U.S. and Canadian Treasury bills mature in six months. Other things remaining constant, what will happen to the exchange rate? Explain.

CHAPTER 11

Practical Hedging Considerations

"Money is always there, but the pockets change."

—Gertrude Stein

This chapter is divided into three sections that concern the everyday practical problems that hedgers face. The first section covers how hedging is classified, the second discusses cross hedging, and the third is on selective hedging.

Hedging Classifications

Hedging is classified into two broad areas: situational and full risk. Each of these two areas is determined by the type of risk and how it is hedged.

Situational Hedging

When a particular price risk in the cash market has been determined and properly hedged, then an individual price risk situation has been resolved. Thus, when a price risk situation has been evaluated and hedged but the next price risk situation might be different, then the hedger is involved in situational hedging.

Consider, for example, a grain elevator that is also vertically integrated and processes the grains into animal feed. The elevator/processor faces cash price risks of inventory value changes in the elevator operation, forward sells of the grain itself, forward sells of the processed feeds, and inventory value changes in the processing unit. Each of these price risks may occur at different times, require different time holding periods, and will be with different market participants. To properly hedge all of these risks, each cash situation will be unique to the other. Thus, the elevator/processor may have simple inventory hedges, forward sell hedges, spread hedges, and basis contracts. In other words, the operation does not have one major type of price risk to handle, but several. Each requires a unique situational evaluation. It is difficult for this type of operation to hedge all their price risks all the time.

Situational hedging lends itself to selective hedging as well. Because each cash price situation has to be evaluated on how to hedge, a natural extension is to add a price forecasting tool and selectively hedge.

Full-Risk Hedging

When an operation faces only one major type of price risk, they can generally cover all their price risk all the time. Grain, bond, and other commodities dealers fall into this category. They face holding period risk. They either forward buy and later sell or forward sell and later buy. They therefore are either long hedged or short hedged, generally for only brief periods of time. If they simply hedge the positions, they have basis risk. However, if they basis trade, they remove a significant portion of basis risk and thus, for all practical purposes, have their full (price) risk hedged.

They do not buy/sell or sell/buy in terms of price, but rather in terms of basis. They try to structure an initial position and a final position at a favorable basis difference. Their revenue margin is the difference between the two basis values. All trades are hedged, and consequently all their price risk is hedged fully all the time.

Full-risk hedgers are margin managers in that they accept small margins on trades but trade large volumes. They can accept small margins because of the volume of trades and because they have totally controlled their price risk.

Obviously large integrated firms can be both full-risk hedgers and situational hedgers. If they can identify their major price risk and if it's fairly uniform, then they can become full-risk basis hedgers for that division. The other price risks can be handled on a situation-by-situation case.

Cross Hedging

The major problem with hedging any commodity is that the cash position and the futures position do not match exactly. Because futures contracts are standardized, the terms of the contract often do not fit the cash position of the hedger. This nonalignment can take several forms:

1. Different Commodities—the futures contract is for one type of commodity, and the cash position or need is for another similar commodity, such as the futures on Treasury bonds and a cash position in AA Corporate Bonds or the futures contract for corn and a cash position in alfalfa.

2. Different Grades/Standards/Maturities—the futures is for only a specific grade or standard and/or maturity, and the cash position is for another, such as the futures contract is specified as #2 yellow corn grade and the cash position is #3 grade, or the futures is for delivery of an average maturity of 20-year T-bonds with an adjusted coupon of 8 percent and the cash position is 15-year 9¼ percent coupon T-bonds.

3. Different Time Periods—the cash position must be entered or liquidated on a time schedule other than the specified futures delivery periods, such as a cash position that is expected to be liquidated on February 15 and the futures is for a March delivery period.

4. Different Quantities—the futures contract standardized size units do not match the cash position amount such as the futures calls for 40,000 pounds of fed cattle and the cash position is 30,000 pounds.

Any one or a combination of all four of these conditions will cause problems for the hedger, mainly through basis variability or delivery against the futures, should delivery be necessary or possible. The difference in quantity will also require a different calculation of net hedge price to reflect over- or underhedging.

The first three of these considerations can be appropriately called cross hedging problems. Cross hedging is simply the process of hedging when the cash and futures positions do not match exactly. The fourth condition, one of differences in quantity, is also a cross hedging problem but is more correctly called the problem of over- or underhedging.

Let's consider each of the four major problem areas and discuss ways to practically handle the situations.

Different Commodities

This issue ranges from closely related and similar commodities to those that are seemingly unrelated. Is hedging possible for a firm that has short-term nonnegotiable certificates of deposit when there is no futures contract that exactly matches the cash position? Is hedging possible for a rancher who has all heifers and the feeder cattle futures contract calls for steers? These two situations are similar in that there are futures contracts that have terms that closely resemble the cash position. What about hedging a cash alfalfa hay position with a corn futures contract, a #2 grain sorghum position with a corn futures contract, or a fluid milk position with a cheddar cheese? Although each of these products is closely related, they differ considerably in physical makeup and market structures. The real issue is not so much that the cash and futures commodities are closely related but that their price movements are closely related.

Hedging is a process that attempts to provide a gain in the futures position that would offset a corresponding loss in the cash position. For this to happen, the price movements of the cash position and the futures position must be very similar. The extent to which they are not results in basis risk; obviously, when the cash and futures commodities are not the same—basis risk.

The real issue of hedging and cross hedging is price movements. To compare price movements, simple graphical analysis can provide useful information, as can simple correlation analysis.

Hedging and cross hedging should only be attempted if the price movements are similar and the basis risk is acceptable to the hedger. Furthermore, just because the futures and cash positions are in the same commodity does not mean that the price movements will be similar consistently. Research shows that during certain periods of time the cash and futures price movements do not have a high degree of correlation, even for commodities for which the cash and futures positions are the same.

Before hedging or cross hedging, the hedger must study the historical relationship between the cash price and the proposed futures price to determine whether they are correlated. Check the time period that the hedge will be placed. The cash and futures may have a high degree of correlation in general, but for a specific time period, especially if it is a short time period, they may not have a very high correlation. For this reason, it is very useful, and approaching necessary, to keep price charts.

Different Grades/Standards/Maturities

A cash position may be composed of USDA #3 corn, yet the corn futures calls for USDA #2 yellow corn. Maturities may differ for T-bonds between the cash and futures. Again, the process is to study the historical price relationships to determine the degree of correlation between the cash and futures. At first blush, it would seem that if they only differ by one grade level, as in the corn example, the price movements would be very similar. Evidence suggests, however, that hedgers should be very cautious when attempting to cross hedge different grades.

Each grade of a commodity has specific uses, thus different supply and demand factors that can cause increased basis variability. The best starting point to determine if cross hedging should be attempted is a detailed analysis of the historical price relationships.

Different Time Periods

Hedges and cross hedges should, for beginning hedgers at least, be placed using a futures contract month that expires as close to the actual final cash position as possible, but not before. If placed before the final cash position, then the hedge will have to be lifted before the final transaction, thus the hedger is no longer hedged and the cash is in a speculative position.

When the final cash position is unknown, then a process of hedging with the nearby, or one contract month away from the nearby, and then rolling the hedge is a common and accepted hedging practice. Hedging with close-by futures maturity months keeps the basis more predictable.

As hedgers become more sophisticated, they may want to consider hedging with a contract month that is not the closest to the actual final cash position. That is, if the final cash position is February 15, the hedge might be placed with a July futures rather than the nearby March contract. This strategy rests on good empirical analysis to evaluate which contract month most closely followed the cash market so that basis changes can be minimized. Similarly, other futures months might provide better basis trading opportunities.

Different Quantities

One of the most troublesome aspects of hedging is the issue of matching the size of the cash position and the futures position. Because futures contracts are standardized by size or quantity, it is difficult to match them exactly with a cash position. A cash position might be $35,000 worth of Treasury bonds, yet the futures contract is denominated at $100,000, or a cash position might be 63,000 pounds of fed cattle but the live cattle futures is for 40,000 pounds at the Chicago Mercantile Exchange and 20,000 pounds at the Mid America Exchange. The dilemma the hedger faces is: Can I hedge, and if so, how much?

Unfortunately there is no easy answer. Because of the inexact amounts of most cash positions, almost 100 percent of all hedges are either overhedged or underhedged. An overhedge occurs when the futures quantity exceeds the cash quantity such that the hedger has a net speculative futures position. A hedger that has a cash fed-cattle position of 63,000 pounds and has it hedged with two live cattle futures contracts (80,000 pounds) is overhedged by 17,000 pounds. The remaining 17,000 pound amount is a speculative futures position. Likewise, underhedging occurs when the cash position exceeds the futures position and the hedger has a net speculative cash position. A bond dealer who had $120,000 Treasury bonds and had them hedged with one T-bond futures ($100,000) is underhedged by $20,000 and thus has $20,000 in a speculative cash position.

The problem can be handled by trying to match quantities as closely as possible. This concept rests upon the premise that if quantities are matched, then the only risk is basis risk. Two broad types of hedges need to be discussed: when the cash position is known and when it is unknown.

Cash Position Known

When the cash position is known, such as with inventory positions, the hedger must try to enter a futures position as close to the quantity of the cash as possible. Consider a cattle dealer who has 66,000 pounds of live cattle. He can sell one Chicago Mercantile Exchange live cattle contract (40,000 pounds) and one Mid America Exchange live cattle contract (20,000 pounds) for a total hedged position of 60,000 pounds. This leaves the trader with a net speculative cash position of only 6,000 pounds. If he purchased two CME contracts (80,000 pounds), he would have a net speculative futures position of 14,000 pounds. Which is best depends entirely upon the direction that price moves. The hedger must calculate the net hedged price properly, as illustrated in Table 11.1. The adjustment multiplies the ratio of the quantities times the profit or loss on the futures position. As shown in Table 11.1, for short hedges, being underhedged is advantageous from a net hedged price standpoint when prices move in favor of the hedger. If prices move against the hedger, being underhedged lowers net hedged price. Being over-hedged produces just the opposite results, as shown in Table 11.2. It must be noted that the examples in these two tables assume no basis changes. Basis changes could mitigate the generalized rules stated here or even reverse them.

Table 11.1 Underhedging Example for Cattle

Cash	Futures
Buy 66,000 pounds of fed cattle at $65.00/cwt.	Sell one CME June contract at $66.00/cwt.
	Sell one MAE June contract at $66.02/cwt.
Price Decrease	
Sell 66,000 pounds of fed cattle at $64.00/cwt.	Buy one CME June contract at $64.00/cwt.
– $1.00/cwt.	Buy one MAE June contract at $65.02/cwt.
	CME (40,000 lbs.) + $1.00/cwt.
	MAE (20,000 lbs.) + $1.00/cwt.
	Net $1.00/cwt. (60,000)
Net Loss = $660.00	Net Gain = $600.00

$$\text{Net Hedged Price} = \$64.00 + \$1.00\left(\frac{60{,}000}{66{,}000}\right) = \$64.00 + \$.91 = \$64.91/\text{cwt.}$$

Table 11.1 Underhedging Example for Cattle *(concluded)*

Cash	Futures
Price Increase	
Sell 66,000 pounds of fed cattle at	Buy one CME June contract at $67.00/cwt.
$66.00/cwt.	
+ $1.00/cwt.	Buy one MAE June contract at $67.02/cwt.
	CME (40,000) – $1.00/cwt.
	MAE (20,000) – $1.00/cwt.

$$\text{Net Hedged Price} = \$66.00 + \$1.00\left(\frac{60{,}000}{66{,}000}\right) = \$66.00 + \$.91 = \$66.91/\text{cwt.}$$

Table 11.2 Overhedging Example for Cattle

Cash	Futures
May 1	
Buy 66,000 pounds of fed cattle at $65.00/cwt.	Sell two CME June contracts at $66.00/cwt.
Price Decrease	
May 5	
Sell 66,000 pounds of fed cattle at	Buy two CME June contracts at $64.00/cwt.
$64.00/cwt.	$65.00/cwt.
– $1.00/cwt.	+ $1.00/cwt.
	$1.00 × 400 × 2 = $800.00
Net Loss = $660.00	Net Gain = $800.00

$$\text{Net Hedged Price} = \$64.00 + \$1.00\left(\frac{80{,}000}{66{,}000}\right) = \$64.00 + \$1.21 = \$65.21/\text{cwt.}$$

Cash	Futures
Price Increase	
May 5	
Sell 66,000 pounds of fed cattle at	Buy two CME June contracts at $67.00/cwt.
$66.00/cwt.	$67.02/cwt.
+ $1.00/cwt.	– $1.00/cwt.
Net Gain = $660.00	Net Loss = $800.00

$$\text{Net Hedged Price} = \$66.00 + \$1.00\left(\frac{80{,}000}{66{,}000}\right) = \$66.00 - \$1.21 = \$64.79/\text{cwt.}$$

The easiest way to handle the problem is to use a combination of regular size futures and minicontracts to reach a futures position as close as possible to the cash position and remain either slightly underhedged or slightly overhedged. If minicontracts are not available, then use regular size contracts and get as close as possible to the cash position and either be slightly underhedged or slightly overhedged.

Alternatively the hedger can attempt to determine price and basis movements and selectively hedge based upon the best forecasts available. This certainly involves more work, and may or may not prove profitable, depending mostly upon the accuracy of the forecasts.

Cash Position Unknown

For producers of products, especially agricultural products (because final yields are dependent on weather and other biological events), the final amount that will be available for sale is uncertain. Thus, added to the problems discussed concerning hedging when the cash position is known is the problem of not knowing the size of the cash position. Consider, for example, a corn producer who estimates that a crop will yield 20,000 bushels. She could hedge with four corn futures contracts and have quantity equivalency. However, if poor weather or disease cause yield loss, she will be overhedged, or if exceptionally good weather occurs, she could be underhedged. This risk, however, can be handled in a way that minimizes the amount of uncertainty concerning production.

First, the hedger needs to have good historical production records. Information can also be obtained from the Cooperative Extension Service using soil test and variety information. The producer can thus make a good estimate concerning potential yields. The hedge should be placed using this estimate.

As the crop grows, the producer can and should adjust the estimated yield as weather patterns and other factors unfold. USDA/State Department of Agriculture Crop/Weather reports are good sources of local information, as are experience and the professional judgment of others in revising yield estimates. The hedge can be increased or decreased (scaled up or down) based upon the revised estimates. This process should be undertaken on a regular basis throughout the growing season and as harvest begins. An example is illustrated in Table 11.3.

Table 11.3 Adjusting a Production Hedge

Cash	Futures
May 1	
Estimates production in November will be 15,000 bushels	Sell three December corn contracts at $2.50/bushel
July 15	
Survey by local Extension Agronomist indicates a reduced yield to approximately 11,000 bushels	Buy one December corn contract at $2.43/bushel $.07/bushel for 5,000 bushels

Table 11.3 Adjusting a Production Hedge *(concluded)*

Cash	Futures
November 15	
Harvest and sell 9,500 bushels at $2.00/bushel	Buy two December corn contracts at $2.10/bushel

$$\text{Net Hedged Price} = \$2.00 + \$.40\left(\frac{10{,}000}{9{,}500}\right) + \$.07\left(\frac{5{,}000}{9{,}500}\right) = \$2.46$$

Minimum Risk Hedges

An alternative to trying to match quantities as close as possible is to consider a minimum risk hedge. A minimum risk hedge is a hedge that tries to equate dollar movements rather than quantity amounts. The basic idea is that the hedger wants price protection from the hedge, not profit. True price protection occurs when the cash and futures move by exactly the same and there is no basis change. Obviously this rarely occurs. Therefore, some other process must be used to achieve dollar equivalency rather than matching quantities. One way to do this is to mismatch quantities based upon historical information. The mismatching of the quantities attempts to eliminate the effects of both price movements and basis movements.

In the example in Table 11.4, the cash position has 25,000 bushels of corn to protect, but the hedging ratio is 1.6, or 40,000 bushels (8 contracts) instead of 25,000 bushels (5 contracts). The example shows a basis movement against the hedger (from $.15 to $.18 for a change of $.03). If the hedger had placed a typical hedge, the results would be a $750.00 loss ($.03 × 25,000 bushels). However, the cash position had an $.08 loss per bushel ($2,000 total), but the overhedged futures position had a gain of $.05 per bushel on 40,000 bushels or a gain of $2,000 to offset the cash loss of $2,000. Thus a minimum risk hedge has occurred. The basis loss was compensated for by an overhedged futures position.

Table 11.4 Hedging Ratio Example

Cash	Futures	Basis
November 1		
Buys 25,000 bushels of corn at $2.50/bu.	Sells 40,000 bushels (eight contracts) of December corn at $2.65/bu.	$.15
December 1		
Sells 25,000 bushels of corn at	Buys 40,000 bushels (eight contracts) of December corn at	
$2.42	$2.60	.18
– $.08	+ $.05	change of $.03
Loss = 25,000 × $.08 = $2,000	Gain = 40,000 × $.05 = $2,000	

To do this requires some empirical work. If done properly, not only can the risk-minimizing hedging ratio be determined, but a range of hedging ratios can be calculated such that the hedger can choose a hedging ratio that minimizes risk or one that increases the hedged returns but has a higher level of risk.

Hedging ratios must be calculated using past information and are at best a guide to what might happen in the future. Hedging ratios are most beneficial to hedgers that have stable and somewhat predictable basis patterns. Also, if used fully, they eliminate basis gains. Consequently, hedgers using hedging ratios to help eliminate the effect of adverse basis movements may want to selectively hedge so that they can take advantage of basis movements in their favor.

Selective Hedging

A naive hedge is a hedge that is placed when a cash position is assumed and the hedge is maintained until the cash position is liquidated. As hedgers become more comfortable with the process of hedging and are no longer unnerved when they receive margin calls, they start to think about ways of having their cake and eating it too. That is, they want to find ways to be hedged when cash market prices are moving against them and to not be hedged when cash prices are moving in their favor. The process of trying to do this is called selective hedging.

Selective hedging has been shown to be effective in increasing net hedged prices. To be effective, selective hedging must be used in conjunction with forecasting methods or trading strategies. There is an infinite number of possible trading strategies for selective hedging. What hedgers want is to be hedged when they need to be and not otherwise—that requires a strategy and a forecast.

Consider a gold dealer who has gold in inventory. She decides to hedge the inventory against the possibility of a price decrease. She wants to be hedged if cash prices decrease, but if prices increase she would prefer not to be hedged. What she needs is a forecast and/or strategy that accomplishes this goal. For example, if she has the gold purchased at $450 per ounce and hedged with a December gold futures at $460 per ounce and prices decrease to $440 in the cash and $450 in the futures, she is protected. But if they rebound back to $460 in the futures and then continue upward, she is prevented from making a profit and furthermore she must make margin calls. The idea of selective hedging is to be unhedged when cash prices start moving up. If prices increased to $480 and then started down again, she could replace the hedge and be protected. To do this effectively requires a plan, strategy, and forecast. Otherwise, the double whammy is possible and losses can double. Table 11.5 illustrates the process of a selective hedge, and Table 11.6 shows the research results of testing several trading strategies. The research shows that a simple hedge combining steers and heifers had a loss of $21.20 per head, but a simple selective hedge with a $3.00/cwt. trading limit reduced the loss to $13.00 per head. In other words, selective hedging does not have a great deal of potential.

Table 11.5 Example of a Selective Hedge

Cash	Futures
May 1	
Buys gold at $450/oz.	Sells December gold at $460/oz.
May 15	
Prices are at $440/oz. but forecasted to increase	Prices at $450/oz.
	Hedge lifted at $451/oz.
	+ $9/oz.
May 15–October 1	
Prices continue slowly upward to $470 but forecasted to decrease	Prices at $480/oz.
	Hedge placed at $479/oz., that is, Sell December at $479/oz.
October 15	
Sells gold at $460/oz.	Sells December gold at $470/oz.
	+ $9/oz.
Net Hedged Price = $460 + $9 + $9 = $478/oz.	

The growth in personal computers during the 1980s and 1990s has increased the use of selective hedging. Software abounds with various strategies for hedgers that gives the place/lift time for investors to use selective hedging.

Hedging Revisited

A final word about hedging is in order. Chapters 1 and 4 emphasized the importance of determining whether an individual will hedge or speculate. This is among the easiest answers to arrive at initially and the most difficult to maintain. Hedging, even in its purest form, is also a type of speculating. Basis movements are a natural part of hedging. Similarly, the issues raised in this chapter are additional risks that hedging entails. Hedging does not eliminate risks—it only helps minimize existing risks.

If you have determined that you are a hedger, then stick to the decision. Develop a trading plan, price objective, and strategy to achieve the plan—and stick to the plan. If you abandoned the plan as soon as you get a margin call, you have increased the amount and kinds of risks with which you must deal. If you have determined that a proper hedge will yield a price objective and you are happy with that price, then don't change your plan when you see that you can get a higher price by dropping the plan. The purpose of a plan is to minimize the greed factor of hedgers.

Table 11.6 Selective Hedging Research Results

Average Profit or Loss per Head, By Sex, From Both Cash and Futures Markets, with Specified Hedging Strategies for Feedlot Operators, New Mexico/West Texas, June 1, 1971 to January 3, 1977

Strategy	Profit Target Limits ($/cwt.)	Steers and Heifers		Steers		Heifers	
		Futures	Combined[a]	Futures	Combined[a]	Futures	Combined[a]
		($/head)					
1. No-hedge[b]	—	N/A	–24.50	N/A	–25.10	N/A	–23.60
2. Routine hedge	—	4.30	–21.20	–1.40	–26.50	12.20	–11.40
3. Selective hedge	3.00	11.50	–13.00	11.30	–13.80	11.70	–11.90
	2.50	10.70	–13.70	10.00	–15.10	11.70	–11.90
	2.00	10.10	–14.40	8.00	–16.40	12.10	–11.50
	1.50	10.30	–14.20	9.00	–16.10	12.00	–11.60
4. Moving averages	5.00 and 3.00	12.30	–12.20	12.30	–12.80	12.30	–11.30
3- and 10-day	4.00 and 2.50	11.90	–12.60	14.10	–11.00	8.80	–14.80
	3.00 and 2.00	11.00	–13.50	11.70	–13.40	9.90	–13.60
	2.00 and 1.50	11.10	–13.40	11.40	–13.70	10.70	–12.90
4- and 18-day	3.00 and 2.00	10.00	–14.50	9.30	–15.80	10.80	–12.80
5. Tolerance intervals	5.00 and 3.00	7.20	–17.30	6.50	–18.60	8.10	–15.50

a. Combined profit or loss from the futures and the cash markets.
b. Only cash market profit and losses.

Hedging is not necessarily something that should be done all the time. Consider each cash transaction separately, determine the risk of remaining unhedged, and the risks of hedging and then make the decision concerning whether or not to hedge. If you are a novice, then hedge on paper first. Keep records and paper hedge as if actually hedging and learn from the results. After numerous paper trades, then move to limited real hedging, perhaps with mini-contracts, and gradually develop a personal hedging program suited to your own individual needs.

PROBLEMS

1. Katrina has purchased 9,000 bushels of corn in the cash market for $3.56 per bushel. She hedges them with two December corn futures (5,000 bushels each) at a price of $3.76 per bushel. Katrina later sells the cash corn for $3.50 per bushel and lifts the hedge at $3.70 per bushel. What was Katrina's net hedge price? What is the relationship of basis movements to the net price in this problem?
2. What is the difference between full-risk hedgers and minimum-risk hedgers?
3. It is said that cross hedging increases basis risk. Explain.
4. George has purchased a well-diversified portfolio of stocks and hedges it by selling a stock index futures contract. The stock market in general goes up as does the value of George's portfolio. His futures position is, of course, losing value, and he gets his first margin call. George believes that now is the time to try the concept called selective hedging, so he liquidates his hedge. Would you want George to manage your portfolio? Why or why not?

CHAPTER 12

Option Markets: The Basics

"Every morning I get up and look through the Forbes list of the richest people in America. If I'm not there, I go to work."

—Woody Allen

Options have existed for millennia; however, only since the 1980s have they regained the attention of serious investors. Most options that are traded today are options on futures contracts. Thus, they have liquidity and centralized price determination and offer different hedging and speculating opportunities compared to regular futures.

Mechanics of Trading

An **option** is a right but not an obligation to purchase or sell a specific product at a specific price on or before a certain date. Thus, the buyer of an option can choose to buy or sell a product if she wants to, but she is not required to do so. Consider for example a real estate transaction. Mr. Jones has a property that Miss Smith thinks she would like. Miss Smith and Mr. Jones agree on a price of $10,000 and enter an option contract for ninety days. Miss Smith has ninety days to make up her mind whether or not she wants the property for $10,000. Mr. Jones may not sell the property to anyone else for ninety days. If Miss Smith decides she wants the property, she **exercises** the option and buys the property from Mr. Jones for $10,000. If on the other hand she decides she doesn't want the property, she lets the option expire.

Because Mr. Jones cannot sell the property to anyone else during the ninety days and has thus effectively removed the property from the market, he will expect monetary compensation to enter an option contract. The monetary compensation that the buyer (Miss Smith) pays to the seller (Mr. Jones) is called the **premium**. Miss Smith will have to pay Mr. Jones a premium for the option, say $500. The price they agreed to ($10,000) is called the **strike price**.

If the buyer of an option is buying the right to buy a product, the option is called a **call**. If the agreement is for the buyer to sell a product, the option is a called a **put**. When the buyer of an option pays the premium, she receives the right but not the obligation to buy or sell. The seller, by receiving the premium, is obligated to perform.

The previous example between Miss Smith and Mr. Jones was a call option because Miss Smith received the right but not the obligation to purchase the property from Mr. Jones. Mr. Jones was obligated to sell the property to Miss Smith if she exercised the option. He was willing to take on the obligation because he received an adequate premium to compensate him for holding the property off the market for ninety days and for the loss of capital gains if land prices move up. If Miss Smith had let the option expire, Mr. Jones would get to keep the premium. If Miss Smith exercised the option, she would pay Mr. Jones an additional $9,500, because he has the premium of $500, thus completing the $10,000 strike price.

Actual versus Futures

The previous example was an option on the actuals (physicals) because it involved the actual transfer of real property. Real estate and stock options are options on the actuals. Real estate options are privately negotiated and are very individualized for each transaction. Stock options can be individualized, but they also have active centralized trading through exchanges such as the Chicago Board Options Exchange. An active options market on the actuals in metals also exists; however, there is no centralized exchange for metals options or for real estate options.

In addition to options on the actual product, there are also options on futures contracts. A buyer of a call option is buying the right but not the obligation to be long a futures position, whereas a put buyer is buying a right to be short a futures. The options are on standard futures contracts and do not in any way change the underlying futures contract. The option is simply an overlaying contract to buy or sell the futures contract at a specified price within a specified time period and in a specified manner.

The option contracts on the futures are traded in "pits" similar to the futures contracts. Strike prices are set at fixed intervals around the currently trading futures price, and then the premium is traded for each strike price by open outcry auction.

Premiums and Strike Prices

Each exchange is allowed to provide a market for option contracts on any futures contract that they are currently trading. Table 12.1 shows a sample listing of option premiums, strike prices, and commodities from the *Wall Street Journal.* Not all futures contracts have options, nor do all delivery months have options contracts. In the United States, premiums are traded (i.e., the premium is the factor that traders act upon, just as price is traded in futures markets), whereas strike prices are set by the exchange. In the sample quotation page in Table 12.1 the commodity corn has a range of strike prices from $2.30 to $2.80 per bushel. The $2.30 strike price had a settled premium value for the September call of $.10 per bushel, and the December put was at $.09 per bushel for the $2.30 strike price.

The standard practice to determine the strike prices at which options will trade today is to take the settle for the previous day on the futures and at predetermined, fixed intervals set at least three strike prices above and three strike prices below the price that is as close to the **settle price** as possible. The three-above and three-below mechanism provides at least seven available strike prices to trade each day. As the market moves up and down, additional strike prices are added. Each strike price is then available to be traded on the exchange.

Table 12.1 *Wall Street Journal* Sample Quotation Listing for Thursday, July 10, 1997

FUTURES OPTIONS PRICES

Thursday, July 10, 1997

AGRICULTURAL

CORN (CBT)

5,000 bu.; cents per bu

Strike Price	Calls-Settle			Puts-Settle		
	Sep	Dec	Mar	Sep	Dec	Mar
230	10	13	19 1/4	5 1/4	9	7 1/4
240	5 3/4	9 3/8	14 1/4	11	14 3/4	12
250	3 ½	6 3/8	10 ½	18 3/4	22	17 3/4
260	2	4 5/8	7 ½	27 ½	29 1/4	24 3/4
270	1 1/8	3 3/8	5 ½	36 ½	38	32 ½
280	3/4	2 ½	4	46 ½	47	40 3/4

Est vol 17,000 Wd 12,698 calls 11,538 puts
Op Int Wed 176,941 calls 137,180 puts

SOYBEANS (CBT)

5,000 bu.; cents per bu.

Strike Price	Calls-Settle			Puts-Settle		
	Aug	Sep	Nov	Aug	Sep	Nov
725	28 ½	7 7/8	5 1/4	8 1/4	87 ½	132
750	15 1/4	6	4 1/4	19 3/4	110 5/8	156
775	7 3/4	4 ½	3 ½	37 1/8	134	180
800	3 7/8	3 ½	2 3/4	58 1/4	158	204 3/8
825	2 1/8	3	2 ½	81 3/8	182	229 1/4
850	1 1/4	2 3/4	2 1/4	105 ½	206	206

Est vol 220,000 Wd 23,375 calls 9,173 puts
Op int Wed 186,239 calls 88,000 puts

SOYBEAN MEAL (CBT)

100 tons; $ per ton

Strike Price	Calls-Settle			Puts-Settle		
	Aug	Sep	Oct	Aug	Sep	Oct
240	7.25	3.50	2.00	3.50	22.00	.....
250	3.00	2.10	1.50	9.60	30.50	.....
260	1.25	1.50	1.10	17.50	40.00	.....
270	.75	1.00	.70	26.70	.....	.....
280	.25	.75	.....	36.40	.....	.....
290	.15	.50	.....	.....	.....	.....

Est vol 1,650 Wd 1,471 calls 1,799 puts
Op int Wed 23,150 calls 16,799 puts

SOYBEAN OIL (CBT)

60,000 lbs.; cents per lb

Strike Price	Calls-Settle			Puts-Settle		
	Aug	Sep	Oct	Aug	Sep	Oct
2100	.....	.....	.....	.....	.....	.330
2150	.....	.....	.....	.100	.350	.....
2200	.180	.600	.....	.310	.620	.880
2250	.....	.430	.....	.700	.950	1.100
2300	.060	.320	.450	1.150	1.350	.....
2350	.....	.250	.350	1.630	1.780	1.850

Est vol 700 Wd 520 calls 324 puts
Op int Wed 23,895 calls 10,232 puts

Strike Price	Calls-Settle			Puts-Settle		
	Aug	Sep	Oct	Aug	Sep	Oct
150	12.20	15.15	17.45	.45	1.15	1.20
155	8.10	11.25	13.50	1.35	2.25	2.25
160	4.90	7.95	10.00	3.15	3.95	3.80
165	3.05	5.20	7.20	6.30	6.20	5.95
170	1.60	3.25	4.80	9.85	9.25	8.55
175	.75	2.05	3.55	14.00	13.05	12.30

Est vol 0 Wd 25 calls 25 puts
Op int Wed 3,135 calls 2,145 puts

WHEAT (CBT)

5,000 bu.; cents per bu.

Strike Price	Calls-Settle			Puts-Settle		
	Sep	Dec	Mar	Sep	Dec	Mar
300	30 1/4	45 1/4	56	1	3 ½	.....
310	21 3/4	37 3/4	.....	2 3/4	6	9
320	15	31	.....	5 3/4	8 ½	12 1/4
330	10	24 3/4	36 1/4	10 ½	13	16

LIVESTOCK

CATTLE-FEEDER (CME)

50,000 lbs.; cents per lb.

Strike Price	Calls-Settle			Puts-Settle		
	Aug	Sep	Oct	Aug	Sep	Oct
79	2.62	.....	.....	.52	.....	.....
80	1.95	2.22	2.62	.82	1.22	1.35
81	1.35	.....	.....	.....	.....	.....
82	.90	1.20	1.62	1.77	.....	.....
83	.....	.....	.....	.....	.....	.....
84	.40	.....	.87	.....	.....	.....

Est vol 564 Wd 159 calls 260 puts
Op int Wed 4,771 calls 13,370 puts

CATTLE-LIVE (CME)

40,000 lbs.; cents per lb.

Strike Price	Calls-Settle			Puts-Settle		
	Aug	Oct	Dec	Aug	Oct	Dec
63	2.27	4.67	.....	.30	.22	.....
64	1.47	3.80	.....	.50	.20	.25
65	.90	3.05	.....	.92	.42	.....
66	.50	2.30	5.20	1.52	.65	.45
67	.25	1.67	.....	2.27	.90	.....
68	.12	1.12	3.65	3.15	1.25	.85

Est vol 3,221 Wd 937 calls 487 puts
Op int Wed 26,529 calls 24,488 puts

Table 12.1 *Wall Street Journal* Sample Quotation Listing for Thursday, July 10, 1997 *(concluded)*

FUTURES OPTIONS PRICES

Thursday, July 10, 1997

Strike Price	Calls-Settle			Puts-Settle		
	Aug	Sep	Oct	Aug	Sep	Oct
5700	7.40	9.60	.....	4.00	6.20	.....
5750	4.70	7.10	.....	6.30	8.70	8.80
5800	2.90	5.10	.....	9.50	11.60	.....
5850	1.60	3.50	.....	13.10	15.00	.....
5900	1.00	2.50	.....	17.50	19.00	.....

Est vol 4,072 Wd 3,236 calls, 2,796 puts
Op int Wed 27,213 calls 33,851 puts

HOGS-LEAN (CME)
40,000 lbs.; cents per lb.

Strike Price	Calls-Settle			Puts-Settle		
	Jly	Aug	Oct	Jly	Aug	Oct
81	1.92	1.65	.....	.07	1.95	1.95
82	1.10	1.22	.57	.25	2.50	2.50
83	.45	.90	.....	.60	.....	.....
84	.15	.67	.35	1.30	3.92	3.92
85	.05	.45	.....	2.20	.....	.....
86	.....	.35	.....	3.15	.....	.....

Est vol 849 Wd 424 calls 262 puts
Op int Wed 8,101 calls 10,733 puts

METALS

COPPER (CMX)
25,000 lbs.; cents per lb.

Strike Price	Calls-Settle			Puts-Settle		
	Aug	Sep	Oct	Aug	Sep	Oct
100	5.95	6.35	5.85	1.10	2.00	2.95
102	4.35	5.15	4.75	1.50	2.80	3.90
104	3.05	4.00	3.80	2.20	3.65	4.90
106	2.00	3.00	3.00	3.15	4.65	6.10
108	1.30	2.25	2.50	4.45	5.90	7.40
110	.80	1.65	1.90	5.95	7.30	8.85

CANADIAN DOLLAR (CME)
100,000 Can.$, cents per Can.$

Strike Price	Calls-Settle			Puts-Settle		
	Aug	Sep	Oct	Aug	Sep	Oct
7200	.....	12.70	.....	.70	1.60	.....
7250	7.70	8.80	.....	1.50	2.70	.....
7300	4.40	5.90	.....	3.20	4.70	.....
7350	2.40	3.60	.....	.....	7.40	.....
7400	1.20	2.20	.....	.....	10.90	.....
7450	.50	1.20	.....	.....	14.90	.....

Est vol 338 Wd 206 calls 5 puts
Op int Wed 10,437 calls 3,638 puts

BRITISH POUND (CME)
62,500 pounds; cents per pound

Strike Price	Calls-Settle			Puts-Settle		
	Aug	Sep	Oct	Aug	Sep	Oct
16600	3.20	3.74	.....	.72	1.28	.....
16700	2.52	3.12	.....	1.04	....	.....
16800	1.92	2.58	.....	1.44	2.10	.....
16900	1.42	2.06	.....	1.04	.....	.....
17000	1.04	1.64	.....	.....	3.16	.....
17100	.74	1.30	.....	.....	.....	.....

Est vol 1,100 Wd 10,820 calls 9,461 puts
Op int Wed 51,840 calls 53,006 puts

SWISS FRANC (CME)
125,000 francs; cents per franc

Strike Price	Calls-Settle			Puts-Settle		
	Aug	Sep	Oct	Aug	Sep	Oct
6850	14.10	17.00	.....	4.00	7.00	.....
6900	10.90	14.10	.....	5.80	9.00	.....
6950	8.00	11.40	.....	7.90	11.40	.....
7050	.....	.....	.....	.....	.....	.....
7100	.....	.....	.....	....	.....	.....

Est vol 2,544 Wd 577 calls 1,595 puts
Op int Wed 17,866 calls 16,470 puts

BRAZILIAN REAL (CME)
100,000 Braz. reaiz; $ per reais

Strike Price	Calls-Settle			Puts-Settle		
	Aug	Sep	Oct	Aug	Sep	Oct
920	....	....	....	....	....	....

Est vol 0 Wd 0 calls 500 puts
Op int Wed 0 calls 9,722 puts

INTEREST RATE

T-BONDS (CBT)
$100,000; points & 64ths of 100%

Strike Price	Calls-Settle			Puts-Settle		
	Aug	Sep	Dec	Aug	Sep	Dec
112	1-59	2-27	3-01	0-03	0-35	1-34
113	1.05	.....	.....	0-13	.....	.....

Consider as an example December Treasury bonds. If the settle for the previous day on the futures was $88-00, then the array of available strike prices to trade today would be:

	Strike Price
	$91
	90
December T-bond	89
Futures at $88	88
	87
	86
	85

These prices would then be available to be traded, and the value of these strike prices to traders would be what trades, that is, what is the premium for each of these strike prices. The $1 intervals were predetermined by the Chicago Board of Trade and are used every day.

Premium values are influenced by a number of factors, but the major factors are whether the option is a put or a call, the length of time until maturity, the price level of the underlying futures contract, and volatility of the commodity's price. Other things being equal, the longer the time until maturity, the higher the level of premium; and the higher the level of price volatility, the higher the premium level. Traders must have a higher premium for a longer time until maturity because generally, more can happen in a longer time relative to a shorter time period, and, the **time value of money** concepts indicate a lower present value (or higher premium) as time until maturity is extended. Price volatility is risk to the seller because he must perform if the buyer exercises her option; consequently higher price volatility implies more risk to the seller, which in turn indicates that a higher premium is needed. The maturity and volatility components are lumped together into something called **time value**. The difference between the price of the underlying futures and the strike price is called **intrinsic value**.

If the December T-bond futures is trading at $88-00 and the $90-00 strike price for a put option on December T-bond futures is trading at $3-00, what does the $3-00 represent? First, because it is a put option, the buyer is buying the right to be short a December futures at $90-00 when the market is at $88-00. If the buyer immediately exercised the put, he would be put short a December T-bond futures at $90-00 and could buy a December T-bond futures back at $88-00 and have a positive revenue of $2-00. This $2-00 represents the intrinsic value of the option, that is, the buyer must prepay the automatic $2 profit built into (or intrinsic to) this particular trade. The remaining $1-00 of the premium represents time value.

The buyers of options obviously want to obtain the option at the cheapest possible price, and sellers have certain threshold levels that must be covered by the premium. For example, the $2-00 intrinsic value is the absolute minimum premium that a rational seller would accept for a $90-00 strike price put when the market is $88-00. If the option sold for less than $2-00, the buyer would immediately exercise the option and capture any profit. Once the intrinsic value, is established, the seller must determine how much time value he is willing to accept. Similarly, a buyer knows that the premium will have to cover any intrinsic value, so the issue is how much time value is the buyer willing to pay. Thus, a bidding process in the option between buyers and sellers concerning time value is the essence of option trading.

If the underlying futures is trading for $88-00 and a strike price of $86-00 is offered, then for a put the intrinsic value is negative. The buyer of the put at a strike price of $86-00 is buying the right to be short a T-bond futures at $86-00. If the buyer exercised, she would put short at $86-00 and would have to buy back at $88-00, thus incurring a two-dollar loss. When the intrinsic value is negative, the option premium has only time value. The negative intrinsic value is not built into (or does not reduce) the premium because there is no obligation (or rational reason) for the buyer to exercise the option under these circumstances. If the option strike price and the underlying futures are at the same price, then the option has only time value because intrinsic value is zero.

A new set of definitions applies to these options. If the option has a positive intrinsic value, it is called **in-the-money**; if intrinsic value is negative, it is called **out-of-the-money**. If the intrinsic value is zero, the option is called **at-the-money**. Some investors call in-the-money options on-the-money and out-of-the-money options off-the-money. A look at Table 12.2 shows the setup for learning option premiums. Notice that strike prices for puts that are above the underlying futures are all in-the-money and all call strike prices above the underlying futures are out-of-the-money, vice versa for below the underlying futures.

Table 12.2 Relationship between Option Premiums and Strike Prices

		Puts		Calls	
	Strike Prices	Premiums (cents per bushel)			
Underlying December Corn Futures at $3.00/bu.	3.30	.33	in-the-money	.01	out-of-the-money
	3.20	.24	in-the-money	.03	out-of-the-money
	3.10	.15	in-the-money	.04	out-of-the-money
	3.00	.05	at-the-money	.05	at-the-money
	2.90	.04	out-of-the-money	.15	in-the-money
	2.80	.03	out-of-the-money	.24	in-the-money
	2.70	.01	out-of-the-money	.33	in-the-money

Time value is the trader's best guess about the impact of the volatility of the underlying futures price and about the total price movement that will occur before the option expires. Consider an option that has a strike price that is a considerable distance from the underlying futures price and is out-of-the-money (these options are called deep out-of-the-money). Let's say that the underlying futures is at $88-00 (December T-bonds) and the put option strike price is $78-00. Why would a buyer pay anything for an option that is $10-00 out-of-the-money? What could a seller get? It depends on each trader's attitude about what will happen to the market between now and the expiration of the option. What is the probability that the price of December T-bonds will decrease at least $10-00? Is it possible? Probable? If the buyer believes it is highly likely that the market will move at least $10, then he is willing to pay more for the option. Similarly, if the seller believes the same, she will demand a higher price.

Exercising and Retrading Options

Option contracts are just like futures in that they are strong contracts and can thus be retraded. Therefore, for option buyers there are three choices: exercise the option, let the option expire, and sell the option (retrade). Sellers are limited to only two choices: hold the option until the buyer decides to exercise or let the option expire, or buy the option back (retrade).

Buyers, not sellers, determine whether to exercise the option or let it expire. That is, after all, why they were willing to pay a premium. If a buyer exercises the option, they are now placed in a futures position. Once in a futures position, they must post margin and pay another commission. Therefore it costs both an additional commission and the time value of their margin money to exercise the option. Because of the added expense of exercising an option, most buyers choose to retrade the option, that is, sell the option back to the market. But sometimes the market is not liquid enough to allow the retrade, and the option buyer must exercise the option or let it expire.

A seller must wait for the buyer to either exercise or let the option expire, which adds uncertainty to the seller. By allowing options to be retraded, the seller doesn't have to wait. The seller may buy the option back from the market and offset his position in much the same fashion as a long futures may offset with a sell.

Options, like futures, are zero sum. For every buy there is a sell, and for every sell there is a buy. Thus when a buyer of a put retrades by selling a put and offsets his position, the act of selling a put implies a buy on the other side, thus zero summing. There can be no selling unless someone is buying, which emphasizes the need for a liquid market.

Buying Options

A buyer must consider several factors when making the decision to purchase an option. First, the buyer must decide which strike price/premium (i.e., in-the-money, at-the-money, or out-of-the-money), and secondly she must weigh whether or not to exercise, let expire, or retrade the option. The first decision is complex and is handled in more detail in the next chapter. For now, consider the effects to a buyer of exercising or retrading an option.

Miss Jones buys a December Gold call at-the-money at a strike price of $400 for a premium of $10.00 per ounce. Miss Jones is hoping that the price of December gold will increase by more than $10.00 per ounce. Miss Jones had to pay the $10.00 per ounce in advance to her broker in order to buy the option ($1,000.00 total cost—100 ounce contract × $10.00 per ounce). In addition, she will pay a commission to the broker when she offsets the option position, say $.50 per ounce ($50.00 total). For Miss Jones to break even, the price of December gold must go up at least $10.50 per ounce. But this is misleading because it depends on whether Miss Jones exercises or retrades the option. Consider the situation where the price of December gold goes up exactly $10.50 per ounce. Table 12.3 shows the net effect to Miss Jones if she exercises the option. Miss Jones exercises the call and is put long December gold futures at $400.00 per ounce. She then sells the December gold futures at $410.50 and makes $10.50 per ounce in revenue. She subtracts her $10.00 premium and $.50 brokerage fee and has a net of zero. However, because she exercised the option, she had to post margin and pay an extra brokerage fee. If the brokerage fee is an additional $.50 per ounce and the time value of the margin is $.014 per ounce ($5,000.00 margin for one day at 10 percent), then Miss Jones did not break even, she lost an additional $.514 per ounce. Miss Jones needs a price increase of at least $11.014 to break even by exercising the option.

Table 12.3 Example of an Option Buyer Exercising an Option

Miss Jones buys a call on December gold at a strike price of $400.00/oz. for a premium of $10.00/oz.	Current price of December gold futures at $400.00/oz.
Price Increase	
Miss Jones exercises the call option and receives a buy position for the December gold futures at $400.00/oz. She posts the $5,000.00 margin. She then sells the December gold futures at $410.50/oz.	December gold increases to $410.50/oz.
+ $10.50/oz. – 10.00 premium .50 option brokerage fee .50 futures brokerage fee .014 opportunity cost of futures margin Loss of $ .514/oz.	

Miss Jones could retrade and avoid the extra commission and margin requirements of exercising. She would need a movement in the option premium of at least $.50 per ounce to break even. The premium of $10.00 and the $.50 per ounce commission would be covered. If the premium went from $10.00 per ounce for the $400.00 strike price to $10.50, then Miss Jones would break even by retrading.

Notice that in the case of exercising the option, the trader is concerned with what happens to the underlying futures price, whereas the decision to retrade is concerned with the movement in the option premium. The two are not unrelated, but the time value of an option will influence whether or not exercising, letting expire, or retrading is more profitable.

A buyer of a put expects prices to fall, whereas a buyer of a call expects prices to increase. An option buyer can lose no more than the price of the premium, but gains are unlimited. Miss Jones, in the previous example, has limited her losses to the $10.50 per ounce premium plus commissions. She expected the price of December gold to go up, but if it went down to $350.00 per ounce, she would let the option expire and lose her premium and commission cost. As a buyer she has a right but not an obligation. It is possible that the $400.00 strike price might have some time value when the market is at $350.00, if there is still some time left to maturity and/or prices are extremely volatile. If the $400.00 strike price was trading at $.50 per ounce (down from $10.00 per ounce), then Miss Jones could retrade and limit losses to only $10.00 per ounce instead of $10.50 per ounce, thereby recovering a portion of her premium. This is illustrated in Table 12.4.

Table 12.4 Example of an Option Buyer Retrading an Option

Miss Jones buys a call on December gold at a strike price of $400.00/oz. for a December premium of $10.00/oz.	Current price of December gold futures at $400.00/oz.
Price Increase	
$400.00 strike price call premium increases to $10.50/oz. Miss Jones sells the $400.00 strike price call for $10.50/oz. + $.50 – $.50 option brokerage fee 0 net	December gold increases to $410.50/oz.
Price Decrease	
$400.00 strike price call premium decreases to $.50/oz. Miss Jones sells the $400.00 strike price call for $.50/oz. – $9.50	December gold decreases to $350.00/oz.

Selling Options

Sellers of options, because they have an obligation to perform, do not have limited losses and unlimited gains like a buyer. Indeed they have just the opposite—limited gains and unlimited losses. A seller of a put option expects the price of the underlying futures to either be stable (not move much) or go up. A seller of a call expects the price of the underlying futures to either be stable or go down. Why would someone take on the potential for only limited gains for unlimited losses? It simply boils down to levels of risk/reward. The seller obviously should have weighed the impact of these factors relative to the amount of premium that was to be received prior to entering the agreement to sell an option. Sellers of options can earn limited returns if prices stay stable and/or move in a certain direction. Buyers must have a price movement to have the potential of a positive return. This is because options are eroding assets to a buyer. If prices remain stable, then the time component of the time value of the premium starts to erode. Remember, the longer to maturity, the higher the premium. Therefore, as time elapses, the premium decreases, that is, it erodes. Thus, a seller will have a limited return if the price moves in one direction or doesn't move.

Sellers of options have two choices. They can write the option contract covered or naked. If they write the contract covered, then they own a corresponding futures contract. If they write it naked, they do not own a futures position.

Covered Option Writing

If the option is going to be written covered, the writer must own a futures contract. If a put option is written, the writer covers the option by simultaneously selling a futures position. If the buyer exercises the put option, then the seller can deliver the short futures contract to the buyer and thus be covered. However, if the option buyer lets the option expire, then the writer retains ownership of the futures. Table 12.5 shows the effects to the writer of covering a put when the price increases, decreases, and remains the same. The exact opposite would be true for a covered call.

Table 12.5 Example of a Covered Option Writer (Seller)

Option writer sells a put option on December T-bonds at a strike price of \$90-00 for a premium of \$2-00	Covers the transaction by selling a December T-bond futures at \$90-00 posts \$5,000 margin
Price Increase	
Option buyer lets option expire	December T-bonds at \$94-00
Writer has gain of premium amount + \$2-00	Writer offsets futures by buying at \$94-00
	– \$4-00
Net to Writer = – \$2-00	

(continued on the following page)

Table 12.5 Example of a Covered Option Writer (Seller) *(concluded)*

	Price Decrease
Buyer exercises; receives the writers short futures position in the futures	December T-bonds at $86-00
Writer has gain of premium amount + $2-00	Writer has transferred ownership of futures to buyer; thus no futures gain for writer
	Net to Writer = + $2-00
	No Price Change
Buyer lets option expire	December T-bonds at $90-00
Writer has gain of premium amount + $2-00	Writer offsets by buying futures at $90-00
	$0
	Net to Writer = + $2-00

If the buyer exercises the option, then the clearinghouse of the exchange would transfer title to the futures to the buyer from the seller. It doesn't matter at what price the futures position was established. The writer would be responsible for any difference between the actual futures price and the strike price. Any deficit would be paid by the writer to the buyer, and any surplus would be paid back to the writer from the buyer.

When an option is written covered, the premium paid by the buyer to the seller is passed through to the seller immediately. Recognize that the seller of the option, by covering with a futures position, must post margin for the futures position. By passing the premium through to the seller, the seller has a portion of the requirement for the margin.

Selling Options Naked

When the option is written naked, the premium is not passed directly to the seller initially. The clearinghouse of each exchange will hold the premium and treat it similar to a margin for the seller. Because the seller is not covering the option directly by writing it naked, the clearinghouse indirectly covers the option by holding the premium for the seller and making the seller responsible for value losses in the option.

If Mr. Smith writes a put naked at a strike price of $400 for a premium of $10 per ounce on December gold and the price goes down to $380 per ounce, then the option gains in value for the buyer and will be retraded or exercised. Mr. Smith faces a loss as the writer, so the clearinghouse will require Mr. Smith to deposit accumulated losses as they develop, just like a traditional margin call with futures. For puts, the clearinghouse treats naked option writers as if they were long the futures position, and for calls, short the futures position.

In the example above, if the buyer exercised the $400 strike price, the clearinghouse would create a futures contract, putting the buyer short at $400 and also putting the naked writer long. This maintains the zero sum principle of futures. Table 12.6 shows the effects to a naked writer of puts with a price decrease, price increase, and no price change situations.

Table 12.6 Example of Naked Option Writer (Seller)

Option writer sells a put option on December gold at a strike price of $400/oz. for a premium of $10/oz. naked	December gold futures at $400/oz. Writer has no futures position
Price Increase	
Buyer lets the option expire	December gold at $420/oz.
Writer has gain of premium of $10/oz.	
Net to Writer = + $10/oz.	
Price Decrease	
Buyer exercises option	December gold at $390/oz.
Clearing corporation puts the buyer short and puts the naked writer long at $400/oz.	
Writer offsets by selling for a futures loss of $10/oz.	
Net to Writer = $0.00 ($10 futures loss offset by gain of $10 premium)	
No Price Change	
Buyer lets option expire	December gold at $400/oz.
Writer has gain of premium of $10/oz.	
Net to Writer = + $10/oz.	

Table 12.7 compares naked and covered option writing. Notice that the two are mirror images of each other depending upon what price does. It is not easy for an option writer to determine whether or not to write naked or covered; a strategy is required before a writer rationally enters the option process. Writing strategies are developed in the next chapter.

Table 12.7 Comparison between Naked and Covered Option Writing

	Price Increase		Price Decrease		No Price Change	
Writer	**Naked**	**Covered**	**Naked**	**Covered**	**Naked**	**Covered**
Puts	Gains premiums	Unlimited loss potential	Unlimited loss potential	Gains premiums	Gains premiums	Gains premiums
Calls	Unlimited loss potential	Gains premium	Gains premiums	Unlimited loss potential	Gains premiums	Gains premiums

Spreading

Spreading involves strategies that have a trader simultaneously buying and selling options, selling or buying two or more different strike prices for the same delivery date, selling or buying two or more different delivery dates at the same strike price, selling or buying two or more options at different exchanges, and selling or buying two or more options for different commodities. The strategies for spreading are almost limitless and are very similar to spreading strategies for futures.

Spreading with options can involve time differences between option premiums for different delivery dates, from differences between related commodities, space differences between options at different exchanges, and a multitude of different strategies involving the various strike prices. Strategies involving spreading with options in hedging and speculating are developed more fully in the next chapter.

Problems

1. Simone has purchased a call option on December T-bonds at a strike price of $98-00 for a premium of $4-00. The underlying December T-bond futures is trading at $101-00. Explain the time and intrinsic values of the premium.

2. Nelson writes a put option naked. Is he bullish or bearish? What if he wrote it covered? If Nelson is bearish on prices for a particular commodity, what choices does he have with options?

3. If speculators buy options, they have the opportunity to have unlimited gains with the only loss being the premium, no more, no less. But just the opposite is true for option writers regardless of whether or not they write them naked or covered. Why would any speculators write options when they could buy them? Explain.

4. Henry buys a put option on gold at a strike price of $350 per ounce for a premium of $12 per ounce. If Henry's option is in-the-money, what is the approximate price of the underlying futures contract? Explain.

CHAPTER 13

Trading Options

"There is no money in poetry, but then there is no poetry in money, either."

—Robert Graves

Options have proven to be excellent investment tools both for hedgers and for speculators. Hedgers find that options provide a different form of price risk management than futures. Variable strike prices and premiums are popular with speculators and hedgers alike. This chapter concentrates on developing the concepts of options more fully for hedgers and speculators and providing examples of their use.

Hedging

Determining how to use options in hedging is relatively easy. The three tests for hedging with futures apply to options as well: (1) opposite initial cash and futures (options) positions, (2) final cash and initial futures (options) positions the same, and (3) if cash price risk is declining prices, short hedge, and if cash price risk is increasing prices, long hedge. Therefore, whenever the futures position would be a sell, replace with a purchased put, and when the futures position would be a buy, replace with a purchased call. This is illustrated in Table 13.1.

Table 13.1 Hedge Setup for Options Compared to Futures

	Short Hedge	
Cash	**Futures**	**Options**
January 1 Buy 100 ounces of gold at $350/oz.	Sell one February gold futures at $353/oz. ↓ Replace this action with this action. ——→	Buy one call option on February gold futures at a strike price of $355/oz., premium of $5/oz. ↑
	Long Hedge	
Cash	**Futures**	**Options**
January 1 Forward sell gold for delivery in two weeks at $341/oz.	Buy one February gold futures at $340/oz. ↓ Replace this action with this action. ——→	Buy one call option on February gold futures at a strike price of $340/oz., premium of $3/oz. ↑

Theoretical hedging with options involves only buying options, not selling. All hedges in this chapter are set up by initial purchases of options, not initial sells. It is true that certain strategies involving initial selling of options may prove to be valuable from a revenue generating standpoint, but these will be treated as strategies, not hedges. There is some debate among traders about this concept. However, it is generally accepted that the prudent use of options as hedges involves only initial buys. Hopefully the examples in this chapter prove this point.

Short Option Hedging

To begin to illustrate how options can replace futures, at least in the beginning stages of a normal production or inventory hedge, consider a gold dealer who has purchased 100 ounces of cash gold for $435 per ounce. He plans to sell the gold as soon as possible but expects that process to take at least a week. To hedge properly with a futures contract, the gold dealer would sell a nearby gold futures contract. Similarly, to hedge with an option, the dealer would buy a put option on the nearby gold futures contract as shown in Table 13.2. Table 13.2 illustrates what the dealer's net hedge price would be with increasing and decreasing prices. This example shows that the option would either be exercised (if prices decrease) or be allowed to expire (if prices increase) because it would then be worthless. Table 13.3 shows the same example with the option being retraded. In other words, the hedger offsets his option and awards the futures market entirely. Notice that the difference in the net hedged prices between the two examples is due to the time value left in the option premium. In summary, the dealer would exercise the option if prices decrease, because he can then receive a sell position on a February gold futures contract priced at $435 per ounce when the current market for that contract is $415. Conversely, if prices increase, the dealer would not want a short position in February gold futures at $435 per ounce if the current market was at $465.

Table 13.2 Gold Dealer Option Hedge Example

Cash	Option
January 1	
Buys gold at $435/oz.	Buy a put on February gold at a strike price of $435/oz., premium of $5/oz.
Price Decrease	
January 15	
Sells gold at $415/oz.	Exercises the option and receives a sell position on February gold futures at $435/oz.
	Buys February gold futures at $415/oz.
	+ $20/oz.
Net Hedge Selling Price = $415 + $20 – $5 (Premium) = $430/oz.	
Price Increase	
January 15	
Sells gold at $465/oz.	Lets option expire (February gold futures at $465/oz.)
Net Hedge Selling Price = $465 – $5 (Premium) = $460/oz.	

Table 13.3 Gold Dealer Option Hedge Example with Retrading

Cash	Option
January 1	
Buys gold at $435/oz.	Buy a put on February gold at a strike price of $435/oz., premium of $5/oz.
Price Decrease	
January 15	
Sells gold at $415/oz.	$435 strike price put trading at $23/oz.
	Sells put at
	$23
	+ $18/oz.
Net Hedge Selling Price = $415 + $23 – $5 (Premium) = $433/oz.	
Price Increase	
January 15	
Sells gold at $465/oz.	$435 strike price put trading at $1/oz.
	Sells put at $1/oz.
Net Hedge Selling Price = $465 + $1 – $5 (Premium) = $461/oz.	

Long Option Hedging

Long hedgers are worried about cash prices increasing. A proper option hedge is to purchase a call in contrast to a futures hedge to buy a futures contract. If a jewelry manufacturer has forward sold some jewelry for delivery in two months but has yet to purchase the gold necessary to manufacture the jewelry, the manufacturer is concerned that gold prices will increase. The fixed sale price and increased input price would squeeze or eliminate her expected profit margin. This initial forward sell would be hedged with options by buying a call. Table 13.4 illustrates the hedge for both increasing and decreasing gold prices.

To summarize the results of the long hedge example, a jewelry manufacturer wishing to avoid the financial impact of a price increase buys a call option on March gold, which guarantees her the right to buy a March gold futures contract at $400 per ounce, the current cash gold price. If prices increase, the dealer is protected and can either exercise the option to receive a long futures position at $400 when the current market is $425 or she can sell the call. Under either action, the dealer nets a price near $400 per ounce. Conversely, if cash prices fall, the dealer would ignore her call option by letting it expire (or sell it for a small price) and pay the lower cash price.

Table 13.4 Jewelry Manufacturer Option Hedge Example

Cash	Option
January 1	
Forward sells jewelry for delivery in two months based on current gold price of $400.00/oz.	Buys call on March gold at a strike price of $400.00/oz., premium of $3.00/oz.
Price Increase	
February 15	
Buys gold at $425.00/oz. to use in the manufacture of the jewelry	$400.00 strike price call trading at $27.00/oz.
	Sells call at $27.00/oz.
Net Hedge Buying Price = $425.00 – $27.00 + $3.00 = $401.00	
Price Decrease	
February 15	
Buys gold at $380.00/oz. to use in the manufacture of the jewelry	$400.00 strike price call trading at $.50/oz.
	Sells call at $.50/oz.
Net Hedge Buying Price = $380.00 – $.50 + $3.00 = $382.50	

In both the short and long option hedging cases, the hedger is protected against adverse price movements with the option hedge. If the cash position is losing value, the option premium is gaining value and the option can be retraded, or the option can be exercised with positive revenue. However, the nice feature about option hedging, and the feature that attracts many of their users, is that when the cash price moves in favor of the cash position, the option can be allowed to expire or it could be retraded to recover a portion of the initial premium. Option hedging protects against adverse price movements, but allows price movements in favor of the cash position to be captured. Futures hedging, in contrast, protects the hedger from adverse movements but prevents her from receiving the benefits of favorable price movements.

Option Hedging Considerations

To properly hedge with options, a few considerations need to be observed. First, the relationship between the change in the price of the underlying futures and the option premium will impact on the effectiveness of the hedge. Secondly the relationship between the price of the underlying futures and the cash price (i.e., basis) will affect certain hedges.

Deltas

The relationship between the change in the option premium and the price of the underlying futures is called delta (Δ). Mathematically it is expressed as

$$\frac{\textit{Change in the option premium}}{\textit{Change in the price of the futures}} = \textit{Delta}$$

Both the numerator and denominator are expressed in the same cents or dollar units, thus delta generally ranges between zero and one. A delta of .9 means that when the underlying futures price changes by $1.00, the option premium changes by only $.90. Deltas of 1 imply perfect correlation between changes of the futures price and the option premium, and deltas of 0 mean that the option premium did not change when the futures price changed. Deltas express the changes in option premiums as the option moves in- and out-of-the-money and time value changes due to changes in volatility and passage of time.

As a general rule, delta values increase as the option gets deeper in-the-money and approach or become the value of one. As the option moves deeper out-of-the-money, delta approaches or becomes zero. The reasons behind this are fairly straightforward. As an option moves deeper in-the-money, time value decreases to almost nothing and the premium reflects only intrinsic value simply because the option is certain to be exercised. When the option moves deeper out-of-the-money, delta approaches zero because the chance that the option will be exercised becomes more remote. Deltas take on the values of zero or one as the option approaches maturity when the time value is eroded away and the probability of a significant price move is effectively zero.

Deltas can only be calculated after the fact with certainty, but they can serve as a guide for future premium values, if only as an estimate. A delta for yesterday's $450 strike price for December gold can be calculated after the market closes. However, we don't know if that delta will hold for tomorrow. In fact, if the market moves at all, the delta will most certainly change. This does not mean that deltas are not useful. They can serve as guides. The rules of thumb are these:

1. If the underlying futures price remains stable, deltas will decrease in value for those premiums with time value—the more time value in a premium, the more the delta will decline.
2. If the underlying futures price increases, put deltas decrease and call deltas increase.
3. If the underlying futures price decreases, put deltas increase and call deltas decrease.

What do deltas mean for a hedger? Because a delta represents a change in the value of the option premium, deltas can serve as guides for what the hedger can expect as price protection. For example, if a corn dealer has hedged her purchase of cash corn with an option, then what the option premium does reflect is how much protection her cash corn position will have. This is illustrated in Table 13.5. The option gained in value only five cents per bushel, while the cash position decreased in value by ten cents per bushel. The hedge provided only half of the protection needed by the corn dealer. Option hedgers must consider deltas when placing and maintaining their hedges. One way to handle the problem with deltas is with multiples.

Table 13.5 Effects of Delta on Hedging

Cash	Futures
November 1	
Buys corn at $3.00/bushel	Buy one put option on December corn at a strike price of $3.00/bushel and a premium of $.10/bushel.
November 15	
Sells corn at	Sells put at a premium of
$2.90/bushel	$.15/bushel
– $.10/bushel	+ $.05/bushel
Net Loss of $.05/bushel	

Multiples

A multiple is simply a term applied to an options hedge that contains more than one option contract. Multiples are used to achieve dollar equivalency with an option hedge. The formula used is:

$$\frac{1}{delta} = M$$

where

M = number of options contracts necessary to achieve dollar equivalency

Or stated another way, M is the reciprocal of the delta value.

If the delta is .50, then the number of options contracts necessary to achieve dollar equivalency is two. If only one option contract is used to hedge, then the option premium would change by only five cents for every ten-cent change in the underlying futures contract. If two contracts are used to hedge, then the two together will equal the change in the futures price. Effective use of multiples usually requires the hedge to be scaled and sometimes rolled. If the price starts to move, the value of the delta will change and thus the size of the multiple will also change. The hedge will need to be scaled either up or down, as shown in Table 13.6. The delta started at .50 and moves to .95, thus the hedge can be scaled down to just one option contract. Notice the calculation of the net hedged price. If the hedge was maintained at two contracts, the investor would be overhedged by one contract and thus would add another component of speculation to his trades.

Table 13.6 Multiple Hedging Example

Cash	Option
November 1	
Buys corn at $3.00/bushel	Delta = .50 M = 2 Buys two put options on December corn at a strike price of $3.00/bushel; premium of $.10/bushel
November 15	
Cash corn at $2.80/bushel	Delta = .95 $3.00 strike price, premium trading at $.195/bushel Scale down buy Sell one put option at $.195/bushel + $.095/bushel
November 20	
Cash corn at $2.75/bushel	Delta = .95 $3.00 strike price Premium trading at $.2375/bushel $.1375/bushel
Net Hedge Selling Price = $2.75 + $.095 + $.1375 = $2.9825	

Basis

Basis is a factor that remains important to hedgers even though they are using options rather than futures. Basis with options that are exercised is very easy to calculate and becomes the same value as with a futures hedge. In Table 13.7 a hedge for a soybean dealer is shown with both a futures and an option hedge. The beginning basis is different for the futures hedge as compared to the options hedge because the option uses the strike price rather than the actual futures price. The reason is that when the option is exercised, the beginning futures position would be the strike price, not the prevailing futures price. The ending basis is the same for either hedge. The only extra basis consideration that an option hedger needs to consider is the calculation of the beginning basis—if the option is to be exercised. Exercised option hedging will be the same as futures hedging with regards to basis except for the initial calculation of basis using the strike price.

Table 13.7 Impact of Basis on an Option Hedge

Cash	Futures Hedge	Basis	Option Hedge	Basis
October 1				
Buys soybeans at $7.00/bushel	Sells one November futures at $7.18/bushel	.18	Buy one put on November futures at a strike price of $7.20/bushel; premium of $.10/bushel	.20
October 15				
Sells soybeans at $6.80/bushel	Buys one November futures at $6.90 + $.28	.10	Exercises the option and receives a sell on November futures at $7.20/bushel	.10
			Buys November at $6.90 + $.30	
		Δ .08		Δ .10

Net Hedge Selling Price Futures Hedge = $6.80 + $.28 = $7.08
Net Hedge Selling Price Options Hedge = $6.80 + $.30 – $.10 (Premium) = $7.00

For retraded option hedging, basis is more complex and not as easy to see. Basis is the numeric difference between a futures price and a cash price; delta is the relationship between a futures price and the option premium. Then, what is the relationship between the cash price and the option premium? It is both basis and delta. However, the two are not tied together by an explicit theory. What an options hedger who is going to retrade the option needs is not basis or delta so much as the combination of the two. Unfortunately, this is easier to ask for than it is to obtain.

What the hedger needs is not a delta for the options premium and the underlying futures ($delta_F$), but a delta for the options premium and the cash market ($delta_C$).

Basis tables are useful tools to a hedger. Delta$_{FC}$ values are useful tools to an option hedger. However, an option hedger really needs a delta$_{C}$ factor and should consider keeping records that are specific to his or her operation. Notice in Table 13.8 that when delta$_{F}$ is held constant at .90, delta$_{C}$ moves from 1.17 to .96 as the basis changes. Thus what delta$_{C}$ gives the hedge is the relationship between the cash price movements and the option premium. This is precisely what a hedger needs. Thus for option hedgers, knowing delta$_{C}$ is more important than knowing either delta$_{F}$ or basis.

Table 13.8 Delta$_{F}$ and Delta$_{C}$ Relationships

Cash	Futures	Basis	Put Premium	Delta$_{F}$	Delta$_{C}$
$10.00 9.50 Δ .50	$10.20 9.55 Δ .65	$ .20 .05 Δ .15	$ .50 1.085 Δ .585	.9	1.17
$ 9.00 Δ 1.00	$ 9.00 Δ 1.20	$0.00 Δ .20	$ 1.50 Δ 1.00	.9	1.00
$ 8.50 Δ 1.50	$ 8.60 Δ 1.60	$ .10 Δ .10	$ 1.94 Δ 1.44	.9	.96

Selecting an Option

For a hedger with options, the choices are many. Usually at least seven to fifteen different strike prices and premiums are available at any time for the hedger to select. The hedger can select very cheap out-of-the-money options to relatively expensive in-the-money options. Which should a hedger choose? The decision will vary among hedgers and will change for any given hedger as conditions change. It is not an easy decision to make because it involves several factors that must be considered. Only two of the major factors will be discussed—financial requirements and level of price protection.

Financial Requirements

Options, as compared to futures have several different financial requirements. With a futures hedge, often at least 5–10 percent of the value of the contract must be posted as margin. However, the margin will be returned at the closeout of the hedge, with any profits or losses added or subtracted. For an option hedge, a premium must be paid in advance and will be totally lost if the option is allowed to expire. Of course a portion of the premium may be recovered if the option can be retraded.

On the other hand, futures hedgers are subject to margin calls and thus additional cash outlays, after the initial margin is paid. Option hedgers who buy options are not subject to any margin calls thus no additional cash outlays after the initial premium is paid. Of course if a strategy is constructed with options that involve selling options, then the option trader will be subject to margin calls.

Price Protection

When selecting an option, the hedger can choose various levels of price protection. Hedgers may want to select a price level that covers all costs and provides a profit margin, or they might rather choose a price that covers only a portion of costs. These alternative cost/profit levels, as discussed more fully in Chapter 7, might include

1. All costs plus a profit margin
2. All costs
3. Variable costs
4. Fixed costs
5. Cash costs
6. Some percentage of costs

Because at least seven strike prices are usually available to traders, some of these alternative price levels will undoubtedly be available for a hedge. If a trader wants to cover all costs plus a profit margin but a strike price adjusted for the premium would not achieve that level, the trader may have to select a strike price and roll the hedge when (and if) a strike price becomes available that would achieve the desired level.

The process of evaluating option strike prices to determine whether they will meet a certain price level objective is accomplished by taking the strike price and subtracting the premium for short hedgers and adding the premium for long hedgers. Table 13.9 shows this calculation. To meet certain cost-level objectives, the hedger must know all relevant components of his costs (total, variable, fixed, and cash). If a hedger knows his various cost components, then developing price-level objectives is easier and more meaningful.

In determining which price level to choose, the hedger must consider the actual cash outlays for various alternatives. If the hedger wants full cost-level price protection, he may have to pick a deep in-the-money option. This could prove to be more of a cash-flow requirement than his budget would allow. Thus, the decision on what price level of protection the hedger wants is not independent of the financial requirements for a proper hedge.

Table 13.9 Evaluating Option Strike Prices

Short Hedgers

Strike Price
– Premium
Floor Price (minimum price)

Long Hedgers

Strike Price
– Premium
Ceiling Price (maximum price)

Example with a short hedge

	Strike Price (S.P.)	Put Premium (P.P.)	Floor Price (S.P. – P.P.)
	3.50	.35	3.15
	3.40	.27	3.13
Current Futures	3.30	.18	3.12
at $3.20	3.20	.10	3.10
	3.10	.08	3.02
	3.00	.05	2.95
	2.90	.03	2.87

An Option Strategy: Synthetic Option Hedging

Options can be used in various combinations of strategies that will mimic futures hedges. When they correctly mimic a futures hedge, they are called a synthetic hedge. For a synthetic short futures hedge, the option strategy would be the simultaneous buying of put and selling a call at the same strike price.

Table 13.10 shows the effect to the hedger's net hedged price when the price moves both up and down. A synthetic long futures hedge involves simultaneously buying a call and selling a put at the same strike price. This is illustrated in Table 13.11. In both examples it is important to note that the net hedged price of the futures hedge and the synthetic hedge are identical. Some brokers advocate this strategy to hedgers with the following argument: If you buy a put you are bearish, so why not also do the other option trade that is bearish—sell a call. This strategy creates a wash in the option premium—as a buyer you pay a premium, but as a seller you receive the premium. If the trader selects an option that is at or near-the-money, then the premiums will be close to the same and thus will offset each other. Therefore, the trader doesn't have much if any cost of placing the strategy. Notice that the effect of the strategy is the same as that of hedging with a futures contract. With a synthetic hedge you are paying two commissions to the broker, whereas with a futures hedge you pay only one commission. The trader is subject to margin calls with both a futures hedge and a synthetic hedge because he has sold an option. Most synthetic hedges do not offer any advantages over a traditional futures hedge. Some traders develop strategies derived from the basic concept that they believe will give them an edge over a traditional hedge.

Table 13.10 Synthetic Short Hedge

Cash	Futures	Synthetic Hedge with Options	
January 1			
Buy 100 oz. of gold at $400/oz.	Sell one February futures at $410/oz.	Buy put on February gold at S.P. $410/oz., premium $10/oz.	Sell call on February gold at S.P. $410/oz., premium $10/oz.
Price Decrease			
January 15			
Sell 100 oz. of gold at $380/oz.	Buy one February futures at $390/oz. + $20	$410 S.P. put trading at $25 $410 S.P. call trading at $5	
Net Hedged Futures Price = $380 + $20 = $400/oz.			
		Sell put at $25 (net on put $25 – $10 = $15) Buy call at $5 (net on call $10 – $5 = $5) Net = $20	
Net Hedged Synthetic Price = $380 + $20 = $400/oz.			
Price Increase			
January 15			
Sell 100 oz. of gold at $420/oz.	Buy one February futures at $430/oz. – $20	$410 S.P. put trading at $5 $410 S.P. call trading at $25	
Net Hedged Futures Price = $420 – $20 = $400/oz.			
		Sell put at $5 (net on put $5 – $10 = – $5) Buy call at $25 (net on call $10 – $25 = – $15) Net = – $20	
Net Hedged Synthetic Price = $420 – $20 = $400/oz.			

Table 13.11 Synthetic Long Hedge

Cash	Futures	Synthetic Hedge with Options	
January 1			
Forward sell product based on current gold price of $400/oz.	Sell one February futures at $410/oz.	Buy call on February gold at S.P. $410/oz., premium $10/oz.	Sell put on February gold at S.P. $410/oz., premium $10/oz.
Price Decrease			
January 15			
Buy gold at $380/oz.	Buy one February futures at $390/oz. – $20	$410 S.P. call trading at $5 $410 S.P. put trading at $25	
Net Hedged Futures Price = $380 – (– $20) = $400/oz.			
		Sell call at $5 (net on call $5 – $10 = – $5) Buy put at $25 (net on call $10 – $25 = – $15) Net = – $20	
Net Hedged Synthetic Price = $380 – (– $20) = $400/oz.			
Price Increase			
January 15			
Buy gold at $420/oz.	Buy one February futures at $430/oz. + $20	$410 S.P. call trading at $25 $410 S.P. put trading at $5	
Net Hedged Futures Price = $420 – $20 = $400/oz.			
		Sell call at $25 (net on put $25 – $10 = $15) Buy put at $5 (net on call $10 – $5 = $5) Net = $20	
Net Hedged Synthetic Price = $420 – $20 = $400/oz.			

A simple variation of the synthetic short hedge of Table 13.10 is to buy a put and sell a call at different strike prices such that a positive premium is earned up front. The effects of this strategy, accomplished by selecting two strike prices such that the sell premium exceeds the buy premium, are shown in Table 13.12.

Table 13.12 Synthetic Hedge Strategy

Cash	March Futures Price	Synthetic Hedge with Options	
January 1			
Buys 5,000 bushels of corn at $4.00/bushel	Sell at $4.10/bushel	Buy put on March corn at S.P. $4.10/bushel, premium $.10	Sell call on March corn at S.P. $3.90/bushel, premium $.23
Price Decrease		Net Gain in Premium of $.13	
January 15			
Sells 5,000 bushels of corn at $3.80/bushel	Buy at $3.90/bushel + $.20/bushel	$4.10 S.P. put trading at $.32 $3.90 S.P. call trading at $.10	
Net Hedged Futures Price = $3.80 + $.20 = $4.00/bushel			
		Sell put at $.32 (net $.32 – $.10 = $.22) Buy call at $.10 (net $.23 – $.10 = $.13)	
Net Hedged Synthetic Price = $3.80 + $.22 + $.13 = $4.15/bushel			
Price Increase			
January 15			
Sells 5,000 bushels of corn at $4.20/bushel	Buy at $4.30/bushel – $.20/bushel	$4.10 S.P. put trading at $.05 $3.90 S.P. call trading at $.32	
Net Hedged Futures Price = $4.20 – $.20 = $4.00/bushel			
		Sell put at $.05 (net $.05 – $.10 = – $.05) Buy call at $.32 (net $.23 – $.32 = – $.09)	
Net Hedged Synthetic Price = $4.20 – $.05 – $.09 = $4.06/bushel			
Net hedged price of a futures hedge for either a price increase or price decrease = $4.00/bushel			

Price Insurance

Insurance involves the idea that a level of protection is provided when needed for a premium to the buyer. The seller (underwriter) assumes the obligation to provide protection if the buyer needs the protection. Options provide this insurance function with respect to price protection. Some say that futures hedging is a form of insurance. It is not. A properly placed futures hedge provides price protection both when it is needed and when it is not. A buyer of an option can exercise the option or let it expire, thus receiving price protection when it is needed and ignoring the option when the protection is not needed, just as the buyer of an automobile insurance policy exercises his option or insurance policy right to collect for financial damages only if an accident or a theft occurs. The option seller provides the same obligation as does an insurance underwriter (thus the reason option sellers are also called underwriters or simple writers).

Simple option hedging provides true price insurance. The option hedger pays the premium so that she can use the option to provide protection when the protection is needed. When the protection is not needed, the option can be allowed to expire. Clearly, the same is not true for a futures hedger who cannot ignore or reject the hedge when cash price movements are favorable. Both a futures hedge and an option hedge are compared and contrasted in Table 13.13.

Table 13.13 Comparison of a Futures versus an Option Hedge

Cash	Futures	Options
July 1		
Forward sell soybeans at $7.00/bu. for delivery by 7/15	Buy July soybeans at $7.30/bu.	Buy a call on July soybeans at a strike price of $7.30/bu., premium $.10/bu.
	Price Increase	
July 15		
Buy soybeans at $7.30/bu.	Sell July soybeans at $7.60/bu. + $.30	$7.30 strike price call trading at $.40/bu. Sell call at $.35/bu. (net $.35 – $.10 = $.25)
Net Hedge Futures Price = $7.30 – $.30 = $7.00/bu. Net Hedge Options Price = $7.30 – $.25 = $7.05/bu.		
	Price Decrease	
July 15		
Buy soybeans at $6.70/bu.	Sell July soybeans at $7.00/bu. – $.30	$7.30 strike price call trading at $0.00 Let expire Net – $.10 (premium)
Net Hedge Futures Price = $6.70 + $.30 = $7.00/bu. Net Hedge Options Price = $6.70 + $.10 = $6.80/bu.		

It should be clear from Table 13.13 that options provide the function of price insurance. However, another concept is also illustrated in the numeric example in Table 13.13. It is the concept of the theory of second best. When the price moves in favor of the cash position, a trader would prefer to be hedged with an option versus a futures contract, but would really prefer to be unhedged. Thus, an option hedge is second best to being unhedged if price moves in favor of the cash position. However, when prices move against the cash position, the trader would prefer to be hedged with the futures contract versus the option because the option has a premium attached. But both are preferable to being unhedged altogether. The option hedge is second best to the futures when prices move against the cash position. Options as hedges are, therefore, always second best.

This is somewhat of an academic point, however. Because a trader does not know for sure what price will do, knowing that options are always second best doesn't help. At best, the theory of second best helps a trader develop a trading strategy. If the trader strongly believes that the cash price will move in his favor, then the hedge might involve an option rather than a futures. The option hedge perhaps would even be a cheap out-of-the-money strike price. This would limit the cash outlay and provide some protection just in case the trader was wrong.

Speculating

Speculating with options can take on several different forms. Option speculation can be relatively risk free or highly risky. It can be very inexpensive or a good buy (goodbye house, car, savings account). Speculative strategies are unlimited in number and possibilities. Some of the more simple and conventional are presented here.

Buy Strategies

Buying options to speculate limits the financial risk of the speculator to the cost of the premium. No margin calls or additional payments will ever be required. Unless a speculator is irrational, the premium amount is the maximum that can be lost by the trader. Option buyers have limited loss and unlimited gain potential. A bullish speculator would buy a call, and a bearish speculator would buy a put.

If a trader thought that the S&P stock index was going to decrease, she could select a strike price from the available puts that are being traded that day. If the trader picked a $250 strike price at a premium of $4, she would pay a total of $2,000 to speculate that the index will drop at least $4 to cover their premium. If the index went up and stayed up, the most the trader would lose is $2,000 regardless of what price the index achieved. The trader may wish to retrade once the price direction is against her position and recover at least part of her premium.

Option buyers can specify stop loss levels and market-if-touched orders just as futures traders can. Successful traders who speculate with options will buy options and place stop loss orders on the premium level to cut their losses. Table 13.14 shows an example using a speculator who purchased a put. The effects that occur when prices move up, down, and remain stable are shown. If the speculator had set a stop loss at $3, then when prices went up and remained stable, the trader would have been taken out of the market and would have losses of $500 instead of $2,000 and $1,750, as shown in Table 13.14.

Table 13.14 Speculation with Put Options for a Buyer

January 1

Buys put on June S&P Stock Index Futures at S.P. $250.00/unit
Premium $4.00/unit

Current June S&P Index Futures at $251.00/unit

Price Increase

May 1
June S&P Index Futures at $2.80/unit
Put S.P. of $2.50 trading at $0.00
Let Option Expire
Total Loss = $4.00/unit or $2,000.00 (500 units)

Price Decrease

May 1
June S&P Index Futures at $2.30/unit
Put S.P. of $2.50 trading at $23.00/unit
Sell put at $23.00 (net $23.00 – $4.00 = $19.00/unit)
Total Gain = $9,000.00

No Price Change

May 1
June S&P Index Futures at $2.51/unit
Put S.P. $2.50 trading at $.50/unit
Sell put at $.50 (net $.50 – $4.00 = – $3.50)
Total Loss = $1,750.00

Sell Strategies

Sellers of options do not have limited risk, in fact their situation is exactly opposite. They have an obligation to deliver a futures contract at a specified price.

As a result, sellers have limited gain and unlimited loss potential. Option speculators who are bullish or believe that the market will be stable will sell puts; if they are bearish or believe that the market is stable, they will sell calls. Table 13.15 shows a bearish speculator with April live cattle and the effects of a price increase, decrease, and no price change.

The example in Table 13.15 illustrates an effect that is attractive to option sellers—time decay. When prices are stable, the time component of the premium erodes as the option approaches maturity, thus the seller earns the decaying component of the option premium as it erodes. This feature of option premiums allows option writers to capture returns when the price moves a certain direction and/or simply remains stable. Option writers will face revenue loss only when the price moves one direction.

Table 13.15 Speculation with Call Options for a Seller

February 1

Sells April live cattle call (naked) at S.P. $80.00/cwt. Premium $3.00/cwt.	Current April live cattle futures at $81.00/cwt.

Price Increase

March 15

April live cattle futures at $85.00/cwt.

Call S.P. $80.00 trading at $6.00/cwt.

Buy call at $6.00/cwt. (net $3.00 – $6.00 = – $3.00)

Price Decrease

March 15

April live cattle futures at $75.00/cwt.

Call S.P. $80.00 trading at $1.00/cwt.

Buy call at $1.00/cwt. (net $3.00 – $1.00 = $2.00)

No Price Change

March 15

April live cattle futures at $81.00/cwt.

Call S.P. $80.00 trading at $1.20/cwt.

Buy call at $1.20/cwt. (net $3.00 – $1.20 = $1.80)

Buy/Sell Strategies

Most option speculators will, at various times, use a combination of buys and sells. This strategy may involve simple spreading or various combinations of buys/sells and puts/calls.

Spreading

Spreading with options is conceptually the same as spreading with futures in that the idea is to estimate what is normal and when the market deviates from normal. Delta spreading is common and involves simultaneously buying a put and selling a call if the market is expected to decrease and buying a call and selling a put if the market is expected to increase. This type of spread is commonly called a straddle. A numeric example is illustrated in Table 13.16. The idea behind a delta straddle is to capture the increase and decrease in deltas as the market moves.

Table 13.16 Straddle Speculation with Options

Buy put on December gold at S.P. $400.00/oz., premium $1.00/oz. (current December gold at $420.00/oz.)	(Delta .1)
Sell call on December gold at S.P. $420.00/oz., premium $10.00/oz.	(Delta .9)
Price Drops to $390/oz.	
Sell put on December gold at S.P. $400.00/oz., premium $10.50/oz.	(Delta .475)
Buy call on December gold at S.P. $420.00/oz., premium $1.00/oz.	(Delta .45)
Put delta moved from 0.1 to 0.475	
Call delta moved from 0.9 to 0.45	
Put net ($10.50 – $1.00 = $9.50)	
Call net ($10.00 – $1.00 = $9.00)	
Net Total = $18.50/oz.	

Strike price spreading involves buying underpriced strike prices and selling overpriced strike prices. The difficulty of this process, once again, is finding what is normal—that is, what is overpriced and underpriced.

A strangle is a buy/sell strategy that is composed of selling both a deep out-of-the-money put and call simultaneously. The trader hopes the market will not move enough to sufficiently approach each of the strike prices so that gain in intrinsic value will exceed the eroding of time value. This strategy will earn the premiums for the trader if the market doesn't move substantially in either direction. A strangle that works properly is illustrated in Table 13.17.

Table 13.17 Strangle Speculation with Options

	Hypothetical Premiums		
	Strike Prices	Premiums	
		Puts	Calls
	430	35	3
Current December	420	27	5
gold futures at	410	18	8
$400.00/oz.	400	10	10
	390	8	18
	380	5	27
	370	3	35

If market remains stable, as the options approach maturity, the eroding time value will move both premiums towards $0.00, thus the strangle strategy nets $6.00 ($3.00 each)/oz.

Strangle: Sell 379 S.P. put, premium $3.00
Sell 430 S.P. call, premium $3.00

Speculative strategies that involve buys/sells and puts/calls are infinite. Individual traders, as they understand the market more fully, will develop strategies that will capture profit relationships. As profit relationships are developed and traded, the ability to continue to capture profit will diminish as more traders enter the market. Thus, a successful trader is always looking for new profit relationships to develop.

PROBLEMS

1. Susan has 100 ounces of gold in inventory that she purchased for $360 per ounce. She hedges it by buying a March gold put option with a strike price of $360 per ounce and a premium of $10 per ounce. The March futures is currently trading at $360 per ounce. Later Susan sells the cash gold for $365 per ounce. The March futures is trading at $366 per ounce, and the $360 March put option has a premium of $4 per ounce. What is Susan's net hedge price (NHP)? What is the relationship between the NHP, basis, and Susan's overall profit (loss)?
2. If options are always "second best," why do individuals use them?
3. Randy has found what he considers to be the best hedging strategy in the world, because he gets price protection and makes a profit. This is what he

says traders should do: If you are a short hedger, buy a put out-of-the-money by at least two strike prices and simultaneously sell a naked call in-the-money by at least two strike prices (ditto for long hedges with calls). This strategy lets you pay a small premium for the put but receive a larger premium for writing the call, thus a positive cash flow on the hedge. If you are a short hedger and the price moves against your cash position, that is, goes down, then the put you purchased will protect you and the call you wrote won't be exercised. If prices move in favor of your cash position, that is, goes up, then offset both the put and call quickly and remain unhedged because you don't need the protection now. Would you let Randy hedge for you? Why or why not? Explain.

4. Martha has 5,000 bushels of corn purchased at a price of $3.50 per bushel. She wants to hedge with put options on the March corn futures at a strike price of $3.50 per bushel with a premium of $5.00 per bushel and a delta of .5. Set up Martha's t-account. One week later, Martha observes that the delta on the $3.50 strike price is now at .25. Adjust Martha's hedge and explain.

CHAPTER 14

Price Analysis

"Vision is the art of seeing the invisible."

—Jonathan Swift

Most systematic price analysis was developed and applied since the early 1900s. Systematic price analysis follows two distinct forms: Marshallian supply and demand, and technical. Marshallian supply and demand analysis is typically called fundamental price analysis. **Technical analysis** is sometimes referred to as **chart analysis**. Alfred Marshall was the first economist to provide a framework of supply, demand, and price equilibrium that other economists and noneconomists could use to understand the behavior of prices. With the publication by Alfred Marshall in 1890 of *Principles of Economics*, fundamental price analysis became a serious field of study by market participants.

The recognized father of technical price analysis is Charles H. Dow. Dow advanced and used the idea (about the same time that Marshall's theory became popular) that past price movements were an indication of future direction and those movements could be systematically analyzed. Because the early technical work was done by visually charting prices, early technical analysts were called chartists.

Most price analysts were in either camp but rarely admitted to being in both camps until after World War II. During the 1950s and 1960s the random walk or efficient market hypothesis gained new recruits. Serious research was conducted that reinforced the ideas that price changes occur independent of each other (random) and markets effectively discount (reflect) all useful information.

If prices are indeed independent of each other, then technical analysis is worthless; and if markets reflect all useful information, then fundamental analysis is conversely worthless. However, what the random walk research has shown is that prices are independent of each other over a sufficiently long period of time and that the markets reflect the general level of information that market participants have or anticipate. Thus, there is some merit to the idea that price analysis is possible and useful for the short term and useful when it can incorporate information that is not generally known or used. The efficient market hypothesis didn't kill price analysis, but it definitely caused market participants to be cautious and to rethink trading plans.

A moment of reflection by the reader will help put price analysis in perspective. If prices were not random and produced patterns that could be recognized and used, then eventually enough people would discover these patterns and then use that information to cause the patterns to cease to exist. Similarly, if supply and demand factors could be formulated in sufficient strength to produce price predictions that were always accurate, then the act of trading on that information would cause those same models to be wrong!

The sad thing about price analysis is that it is extremely labor- and time-consuming with few tangible rewards. Any individual can be a successful investor in futures and options markets either as a hedger or speculator without knowing anything about price analysis. However, a study of some basics of price analysis will help increase your general understanding of the markets and your confidence as an investor and thus in turn potentially give you an edge in the market place.

Fundamental Analysis

Fundamental analysis is simply the use of supply, demand, and other economic factors to predict prices. This usually involves a good knowledge of economics, statistics, and computer systems. Fundamental analysis can take several forms such as the study of supply and demand factors, seasonal and cyclic factors, and econometrics.

Supply and Demand Analysis

Supply and demand can be studied from the standpoint of observing the factors that affect price and using intuition to determine the relative weights placed upon each factor and then make a guess as to price direction. At first blush this sounds very irrational and unreliable. There is no doubt that it is for most people. However, for someone who has experience and knowledge, it can be both rational and reliable. The human mind is capable of balancing several factors and making a forecast very efficiently. The track record of several individuals who use the "gut feeling" approach is excellent, especially when compared to numerous computer based model forecasts. Most individuals who use the "gut feeling" approach systematically analyze the factors that affect price. They usually form a matrix of the factors and subjectively assign weights. This can be best visualized by doing it on paper, at least initially, as illustrated in Table 14.1.

Table 14.1 Forecasting Using a Matrix of Factors

	Corn	
	Supply	**Demand**
P↑	Weather in Iowa and Nebraska below average by 15 percent	New corn-based plastic announced by USDA
P↓	Price of petroleum is down by 5 percent	Price of wheat is down 21 percent from this time last year

Individuals who analyze price by "gut feeling" and do so with considerable accuracy usually have considerable knowledge about a commodity and experience in the industry. This process is highly individualistic and is unteachable except for basic concepts.

Demand

A consumer derives usefulness (utility) from consuming a unit of a particular good. After the first unit of consumption, the usefulness of each additional unit is less to the consumer. This is called the principle of diminishing marginal utility. Therefore to induce the consumer to consume an additional unit, the maximum price the consumer is willing to pay must be reduced. This produces the classic downward sloping demand curve. Changing the price causes a change in the quantity demanded, as the demand curve remains in place while the price changes to reflect a different quantity.

The relationship between a relative change in the price of a product and the relative change in quantity demanded is called the elasticity of demand with respect to price. This elasticity is expressed as

$$E = -\frac{\frac{\Delta Q}{XQ}}{\frac{\Delta P}{P}} = -\frac{\%\Delta \; in \; Q}{\%\Delta \; in \; P}$$

along the Demand curve

where

E = Elasticity

Q = Quantity

P = Price

When $E = 1$, the elasticity is unitary, that is, when price changes by 1 percent, so does the quantity demanded. If $E < 1$, the demand curve is inelastic. When price changes 1 percent, quantity demanded changes by less than 1 percent. When $E > 1$, then the demand curve is elastic. A 1 percent change in price yields a greater than 1 percent change in quantity demanded.

Although each linear demand curve actually exhibits all three values of E, some demand curves are generally more elastic or inelastic than others. Figure 14.1 shows both an inelastic and an elastic demand curve.

When all the individual demand curves are added together, then a market demand curve is created. Because several factors were held constant to derive a demand curve, when those factors change, the whole demand curve shifts. The major factors that cause a change in demand are price of other related goods (supplements and complements), income, tastes and preferences, and population (number of consumers). Figure 14.2 shows the effect of increasing consumer incomes on the demand for a product. Most products are normal goods, that is, when incomes increase, so does the consumption of the product. However, a few are inferior, and when incomes increase, consumption decreases.

As a recap, the quantity demanded of a product changes when the price of the product changes. However, when either a change in income, tastes and preferences, price of other related goods, or population occurs, the demand curve shifts either right or left.

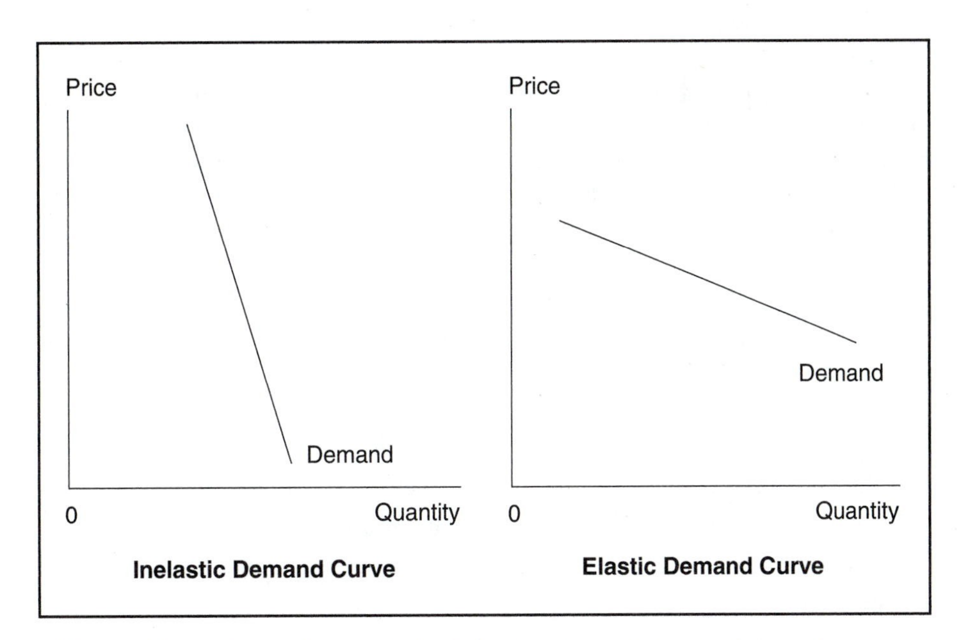

Figure 14.1 Example of relatively inelastic and elastic demand curves

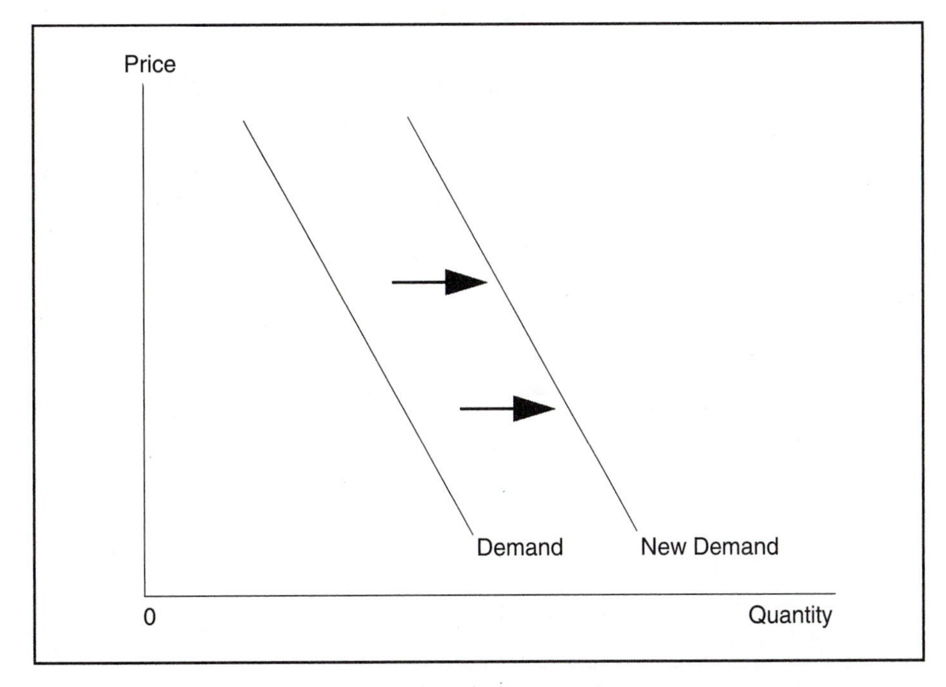

Figure 14.2 Increasing consumer incomes' impact on demand

Supply

Producers of a product will increase the quantity supplied when the price increases and vice versa. Because the supply curve is derived from a firm's cost structure, it is best to think of it as a cost curve. If either unit input prices or technology of production changes, then the supply curve will shift right or left. Figure 14.3 shows the effects of a cost-saving technology on the supply curve.

Equilibrium

The interaction of the two forces of supply and demand yields price. Changes in the equilibrium price will be caused by

Demand	*Supply*
Tastes and preferences	Technology of production
Population (number of consumers)	Input prices
Price of related goods	Population (number of producers)
Income	

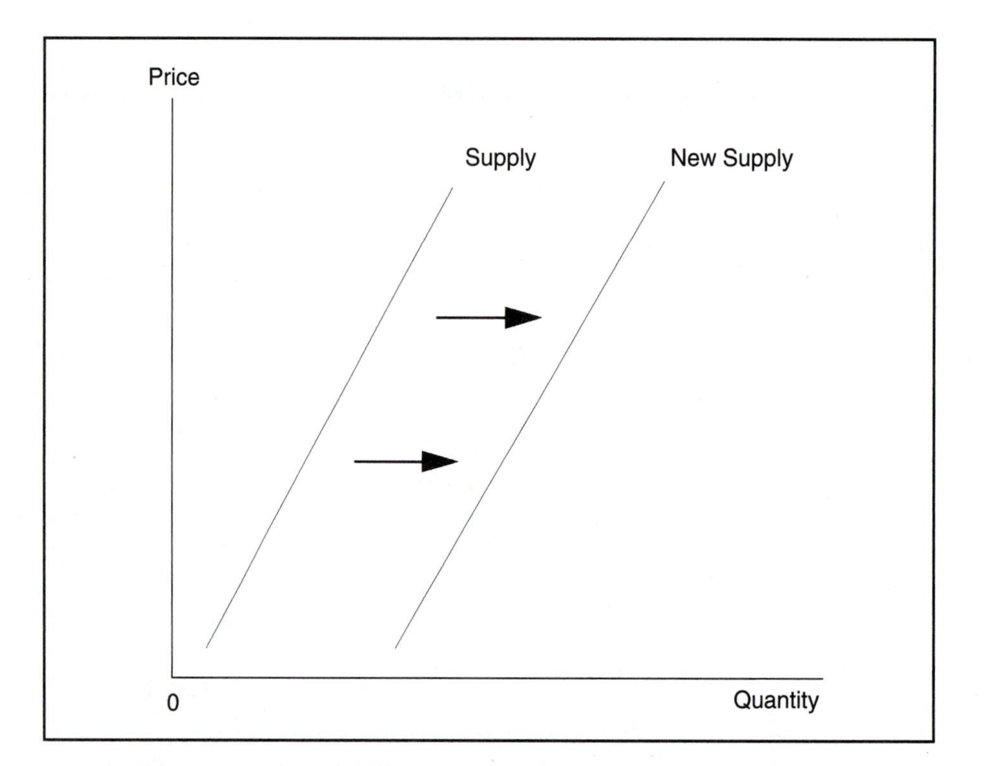

Figure 14.3 Impact of a cost-saving technology on supply

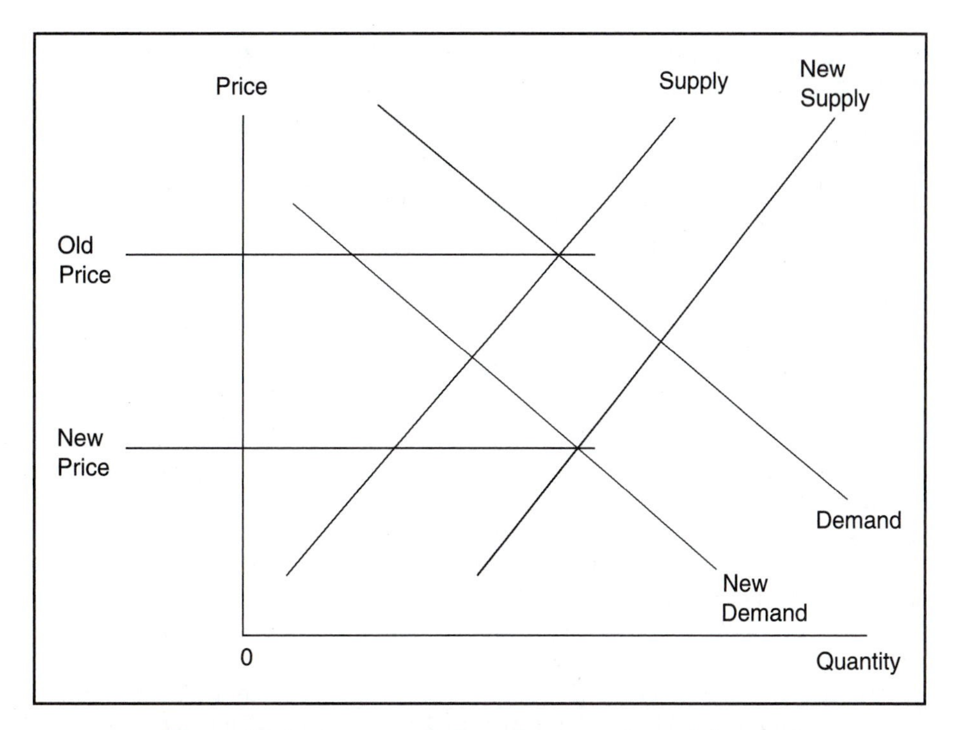

Figure 14.4 Impact of changing supply and demand

Therefore, only seven major factors are responsible for changes in the price of an item. Figure 14.4 shows the effect on price of a change in consumer taste and preferences (negative due to new information about health) and a new cost-saving technology adopted by producers.

The Weighted Matrix Approach

The weighted matrix approach to price analysis simply lists all relevant supply and demand factors that have occurred or will occur. A subjective weight is then placed on each factor. This subjective weight is a relative weight between each supply and demand factor as to which has the strongest impact upon price. Any weighting scale can be used as long as it is consistent and relative. Table 14.2 shows a weighting scale of 1 for the demand factor and 1.2 for the supply factor indicating that the supply factor will have the stronger of the two impacts upon price (20 percent stronger). This particular individual felt that the demand factor would have a positive impact upon price (cotton is a normal good and therefore has a positive income elasticity); however, the increase in production (a negative impact upon price) was felt to be a stronger force upon price. This individual would thus forecast a slight decrease in cotton prices.

Table 14.2 Weighted Matrix of Supply and Demand for Cotton

P↓	Excellent weather conditions in most growing areas indicates a 5 percent increase in production	1.2		
P↑			National incomes projected to be up 2 percent next quarter	1

A variation of the weighted matrix approach uses subjective probabilities of the occurrence of each event. For example, in Table 14.2 each forecasted factor was assumed to have the same weight of occurrence. But if the forecaster felt that the supply forecast had only an 80 percent chance of occurring, then he might alter his relative weight from 1.2 to 1 and thus feel that the price of cotton would remain about the same.

This approach is obviously imprecise and likely inexact. The example showed only two factors impacting upon price when there are usually always several. In fact, it is because there are usually several factors that the matrix approach is appealing. Listing all the factors that you think will impact upon price, the direction price will be impacted, the relative weight each factor should have, and each factor's probability of occurrence will force you to become a better forecaster. It may be inexact, but the approach is consistently better than a simple naive forecast for those who use it, learn from it, and increase their knowledge and experience base for developing later forecasts.

Seasonal and Cyclic Factors

Most commodities will display price movements within the year, often very consistently. Within-year movements are called seasonal movements, and over-several-year movements are called cycles. Cycles are less useful in futures trading because of the time frame. Year-to-year movements are not very helpful for contracts that mature in three months. However, cycles can be helpful in forecasting general strength or weakness in prices for the shorter term. Most traders rely more on seasonals, partly because they are shorter in duration and partly because they are thought to be more accurate than cycles.

Seasonals

The strongest seasonals exist for commodities that have biological or market structures that cause some degree of regularity. Agricultural commodities have strong seasonals mostly because of biological factors. Corn is planted in the spring and harvested in the fall—once a year. Gestation and growth rates affect livestock. Both grain and livestock have regular seasonal patterns, as illustrated in Figures 14.5 and 14.6, respectively. Other commodity futures contracts have less-pronounced seasonals such as Treasury bills or stock indexes. Market structures and conditions such as auction dates and release periods can cause seasonals in financial instruments.

Some seasonal forecasters attempt to discover the causes of the seasonals; other forecasters simply try to determine the pattern and then forecast based upon the pattern. The simple pattern forecasters don't really care what the causes are, they only care what the price direction will be. The seasonal forecasters who

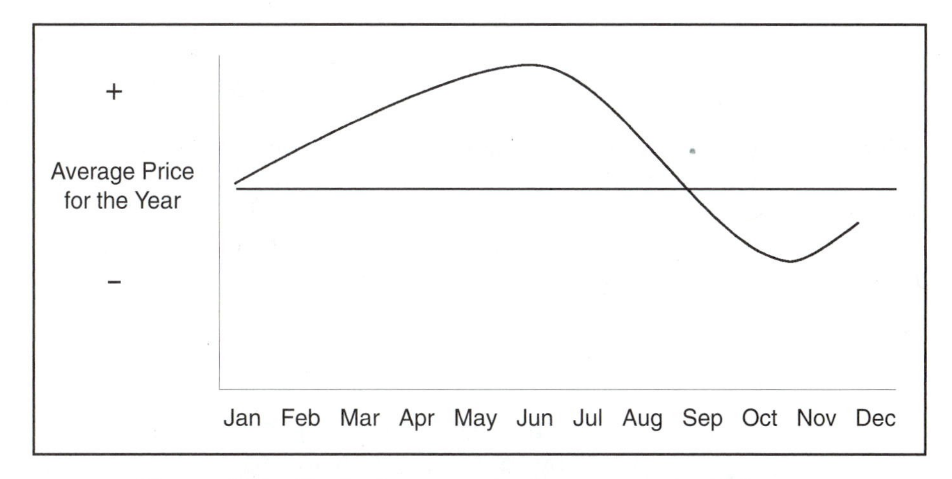

Figure 14.5 Typical seasonal price pattern for U.S. corn

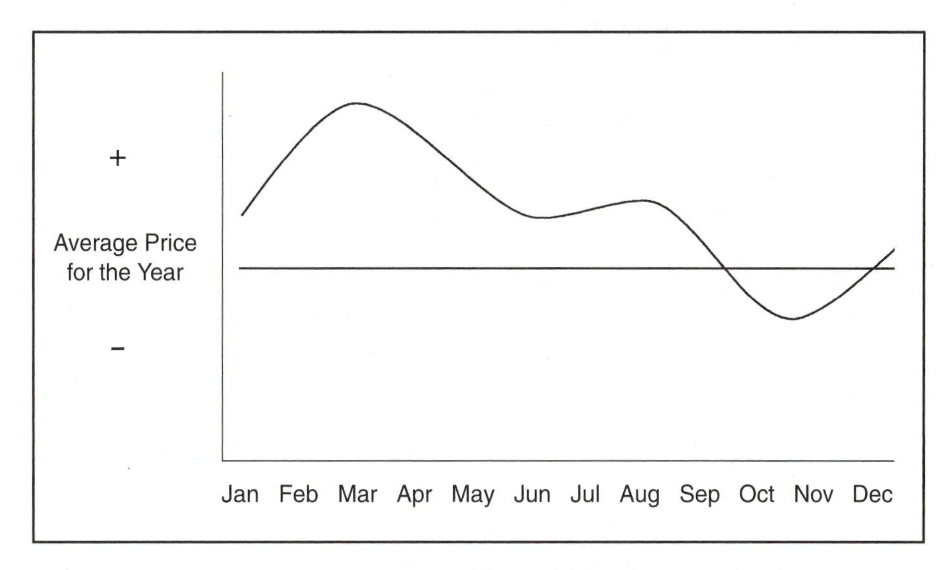

Figure 14.6 Typical seasonal price pattern for U.S. cattle

try to determine why the seasonal price pattern has ups and downs often will use the weighted matrix approach or econometrics to determine what the forecast will be, rather than the simple past pattern of prices.

Cycles

Cycles are more difficult to forecast. Some forecasters simply look for patterns that tend to repeat themselves every few years and forecast from the past patterns. They are aware that these patterns will probably change, but they don't try to determine why. More serious forecasters try to determine the factors that can cause cycles.

The pure academic definition of a cycle is a price pattern that repeats itself with enough regularity to be useful in price forecasting. Using that definition, only livestock are said to have cycles. Cattle and hog cycles go back as far as the late 1860s when records were first kept. Some forecasters found corn cycles, but they faulted after a few ups and downs. Likewise some argue that there are long-term business cycles.

Business and livestock seem to have the only consistent cycles. However, some shorter cyclic periods have occurred, and some forecasters have found that cyclic analysis is useful for many commodities.

Econometrics

The fundamental analysis tool that has enjoyed the most popularity in the last few decades is econometrics. **Econometrics** is nothing more than the application of statistical and mathematical tools to economic relationships so that forecasts can be made because the relationships have been quantified.

A simple price analysis of corn prices might include listing the amount of corn produced versus the price with the thought that the more produced, the lower the price, as illustrated in Table 14.3. However, the tabular presentation doesn't quantify the relationship. The next step might be to graph the numbers in an attempt to draw a "best fit" line through the data. This still is neither a precise nor an unbiased approach. A statistical tool called regression, however, can quantify the relationship and give unbiased and precise results.

Table 14.3 Hypothetical Corn Production and Price

Year	Production (billions of bushels)	Price (dollars per bushel)
1985	7.3	$2.27
1986	8.4	2.16
1987	7.0	2.30
1988	6.9	2.31
1989	8.1	2.19
1990	8.6	2.14

Regression is a process that yields a "best fit" line through the data point that minimizes the sum of the squared deviations between the data point and the regression line. This yields a line that was arrived at in an unbiased way, not like the artistic approach of drawing a line using only the judgment of the artist.

The relationship between corn production and price can be expressed as

$$P = f(Q)$$

where

P = Price of corn

Q = Quantity of corn produced

f = function

If the function is assumed to be linear, then the expression is

$$P = a \pm b(Q)$$

where

a = intercept term

b = slope of the line

Regression will estimate what the parameters a and b are, given the price and quantity data. Using the data in Table 14.3, regression gives an equation of

$$P = 3.0 - .1Q$$

A forecaster can now estimate what the price will be for the next production period when the forecast for the production comes out. If the USDA forecast for next year's production is 8 billion bushels, plugging that number into the formula yields a forecasted price of $2.20 per bushel.

This is the simplest form of econometrics. Serious forecasters will add variables to the equation because they believe that production alone is not the sole determinant of the price of corn. Other factors such as the price of other grains, interest rates, and the price of livestock will all contribute to the price of corn. Also, the functional form may not be linear, it could be a polynomial or logarithmic. Furthermore, only one equation may not be best. Several equations may need to be fit simultaneously.

Econometrics can be very sophisticated involving numerous equations requiring complicated mathematics to solve and elaborate statistical tools to analyze. Unfortunately, there is no good evidence to suggest that the complicated econometric forecasts are consistently any better than the simple models. In fact, the old phrase, KISS (KEEP IT SIMPLE, STUPID), seems to be very appropriate for most would-be forecasters.

There may exist very complex econometric models that are very accurate forecasters, but their creators are keeping them secret. This secretive approach is what any forecaster should do, unless she wants the forecast to be used by everyone and thus become useless (in the sense of being able to use it to make a profit) to all.

Recognize that econometrics is a very valuable tool to validate and to discover underlying economic relationships. Its greatest value lies in applying quantitative rigor to economic concepts. However, the reader must understand that the ability to explain a past economic relationship with rigor does not necessarily mean that it can be applied to future events as a forecast.

Technical Analysis

Technicians believe that past price movements provide an indication of future price movements. A historical record of prices is necessary for technical analysis. Two major tools are used to forecast: charts and mathematical fitting.

Charts

Charts exist in many forms, however two major types are used most often: bar and point and figure.

Bar Charts

A bar chart is composed of the high and low for the day (or week, month, or year) and a mark (bar) where the market settled or closed, as illustrated in Figure 14.7. From the past movements of price, a bar chartist will attempt to forecast the following: whether a trend exists; if so, how long it will continue, and if not, when a trend will form; and where the major turning points are (i.e., when will the up market turn down, the down market turn up, or a sideways market go up or down).

To determine these things, a bar chartist uses several techniques such as trendlines, channels, support, resistance, key reversals, flags, pennants, triangles, gaps, saucers, bowls, double tops, triple tops, double bottoms, triple bottoms, head and shoulders, wedges, hook reversals, and island reversals to name just a few. Several books have been written on the subject of bar chart formations (a few are listed in the recommended reading list [Appendix A]).

Bar chart analysis is very similar to cloud reading—some formations are obvious, but others are seen by only one observer. Successful charting involves art, not science. Thus, people with the ability to spatially see and analyze objects will be the best at "reading" the formations. Even those who have difficulty reading charts find simple bar charts useful to help determine stop loss levels and evaluate relative price comparisons.

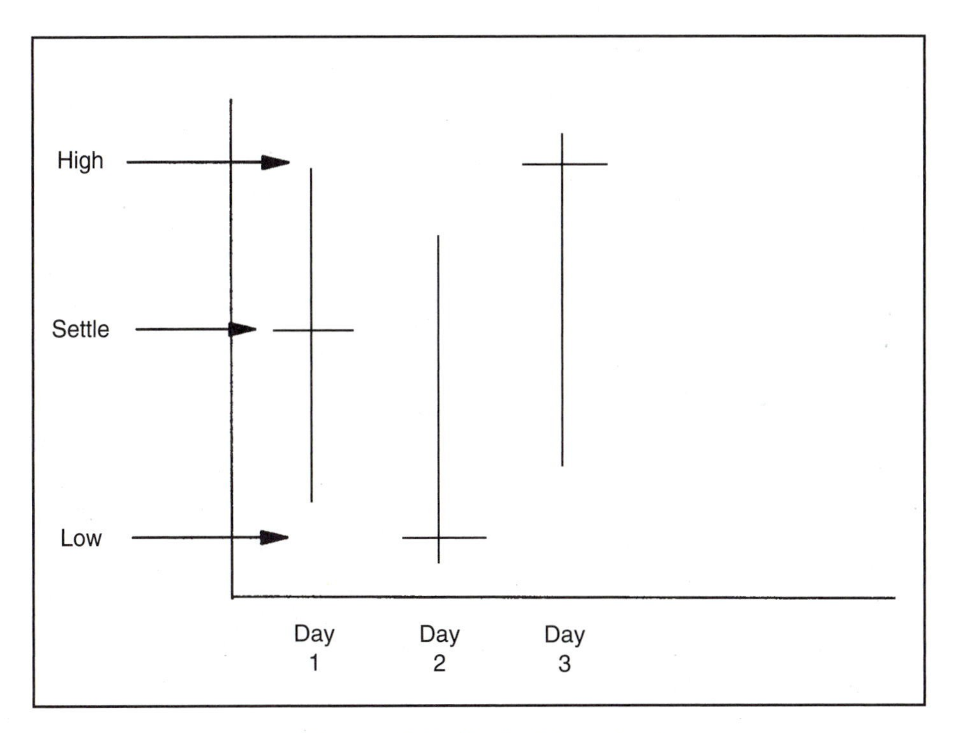

Figure 14.7 Typical bar chart

Point and Figure Charts

Point and figure chartists believe that only market movement is important, thus they eliminate time as a variable. To construct a point and figure chart involves using blocks or cells that represent a certain price range such as five cents. Typically "*X*s" are used to show an up-trend and "*O*s" are used to show a down trend, as illustrated in Figure 14.8. Additionally a reversal requirement is needed to denote how many cells a price must move in order to start a new price direction. Thus, a two-cell reversal would mean that the price must move up or down by two cells (ten cents if each cell represents five cents) to cause a reversal to be charted.

Only the highs and lows are plotted. The settles or closes are not used. Highs are used as long as the market continues up, and lows are used when the market is going down. For a reversal of trend to occur, the price movement must be enough to "drop a cell" plus two cells (the reversal criteria) for a change from up to down. For a change from down to up, the movement must be enough to "add a cell" plus two cells. Therefore, on a 5-x-2 point and figure chart, the price reversal criteria is fifteen cents (drop or add a cell plus two cells, each of which is worth five cents).

One feature of point and figure charts that technicians like over bar charts is that exact buy and sell signals are generated. When the market moves above the previous up level, buy, and when it decreases below the previous low, sell. In

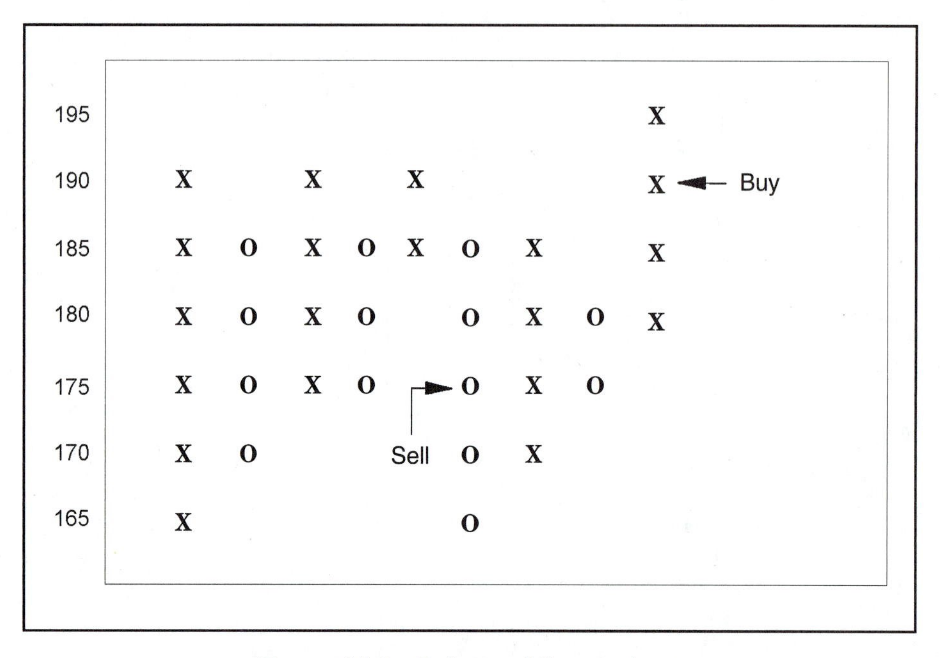

Figure 14.8 Point and figure chart

Figure 14.8, a sell signal occurs when the third-from-left down trend of "*O*s" goes below 180 to 175. The move down continued to 165.

The size of cell selected and the reversal criteria are subjective. If the cell size is large and the reversal criteria large, then only the very major price movements will be recorded. If the cell size is small and the reversal criteria small, then all minor price movements will be recorded. Reading point and figure charts, like bar charts, involves art, not science.

Mathematical Fitting

Many technicians become frustrated with the subjective nature of charts and the interpretation of the various formations and look for a more scientific analysis of price movements. The methods most often chosen include some type of mathematical tools that don't appear to have as high a degree of subjectivity as does reading charts. The two most popular systems are moving averages and wave analysis.

Moving Averages

Moving averages smooth the price information to remove the daily noise of small price moves. The idea is to filter out useless nontrend price movements and concentrate only on major price moves. Usually at least two averages are calculated and often three such as a three-day and nine-day or a three-, nine-, and eighteen-day average. By using two or more different averages, turning point signals can be generated. Table 14.4 shows the calculation of a three- and a nine-day average. Each new day that is added drops a day from the calculation, thus a "moving" average.

Table 14.4 Moving Average Calculation for a 3- and 9-Day Average

Day	Market Close Price	3-Day Average	9-Day Average
1	$2.01		
2	2.03		
3	2.04	2.026[a]	
4	1.99	2.02[b]	
5	1.93	1.986	
6	1.91	1.943	
7	1.89	1.91	
8	1.83	1.876	
9	1.81	1.843	1.937[c]
10	1.79	1.81	1.913[d]
11	1.78	1.793	1.885
12	1.73	1.767	1.851

a. 2.026 = (2.01 + 2.03 + 2.04) ÷ 3
b. 2.020 = (2.03 + 2.04 + 1.99) ÷ 3
c. 1.937 = (2.01 + 2.03 + 2.04 + 1.99 + 1.93 + 1.91 + 1.89 + 1.83 + 1.81) ÷ 9
d. 1.913 = (2.03 + 2.04 + 1.99 + 1.93 + 1.91 + 1.89 + 1.83 + 1.81 + 1.79) ÷ 9

Once the length of the averages is selected, then moving averages, like point and figure chart analysis, give consistent buy and sell signals, as illustrated in Figure 14.9. The signals are very simple and follow a simple rule: When the

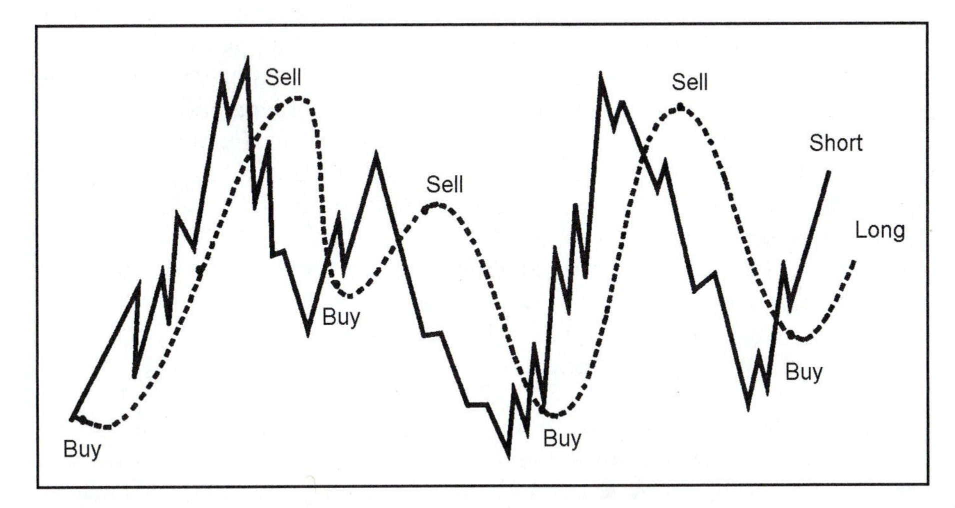

Figure 14.9 Moving average buy and sell signals

short-term average crosses the longer average from above (that is, when the shorter average is less than the longer average), then a sell signal emerges. When the shorter average crosses the longer average from below, a buy signal is generated (when the shorter average becomes greater in value than the longer average).

When three averages are used, the short average crossing the middle (intermediate) average is used as a "watch" signal. When the intermediate average crosses the longest average, buy or sell signals are created.

Moving averages are also "weighted." Some users believe that the most-recent prices should reflect a greater weight; therefore, instead of giving equal weights to each day in a three-day average, the three days are weighted such that the most recent day is most heavily weighted (say 60 percent), the second day less heavily weighted (say 30 percent), and the final day least heavily weighted (say only 10 percent).

The weighting concept is used by technicians to develop their own individual systems. The possibilities are infinite and easily programmed into personal computing software. In fact, many of the systems can be purchased in the form of various software packages and/or books.

Wave Analysis

Moving averages are a form of wave analysis; however, the two are separated because wave analysis takes on a more sophisticated level of mathematics than do averages. Wave analysis includes all mathematical fitting tools other than simple or weighted moving averages.

Some of the simple fitting tools used are the trigonometric sine and cosine functions, polynomials, and logarithms. Past price information is used in conjunction with a sine curve formula or a polynomial equation. A forecast is then possible once the best fit equation is found. This is simply curve fitting as mathematicians call it and has no fundamental reasoning other than finding the best equation to fit the data. This has become an increasingly popular price forecasting tool with the advent and availability of microcomputers.

Wave analysis, however, is usually thought of as the Fibonacci number sequence. The Fibonacci sequence is the list of numbers 1, 1, 2, 3, 5, 8, 13, 21, 34, 55, 89, 144 . . . . Notice that each two numbers that are added together produce the next number in the series ($1 + 1 = 2$; $1 + 2 = 3$; $3 + 2 = 5$; $3 + 5 = 8$, and so on). Additionally, after the first eight numbers, the ratio of the numbers adjacent to each other is either 1.618 or its inverse (.618); the inverse is called phi. Alternate numbers have a ratio of 2.618 or its inverse (.382). Because of the constant ratios, waves appear in price series that mirror a Fibonacci sequence.

R. N. Elliott, in the early part of this century, developed a system called the Elliott Wave Principle that uses Fibonacci numbers. Elliott's theory says that markets move up five waves and down three for a total of eight, as shown in Figure 14.10.

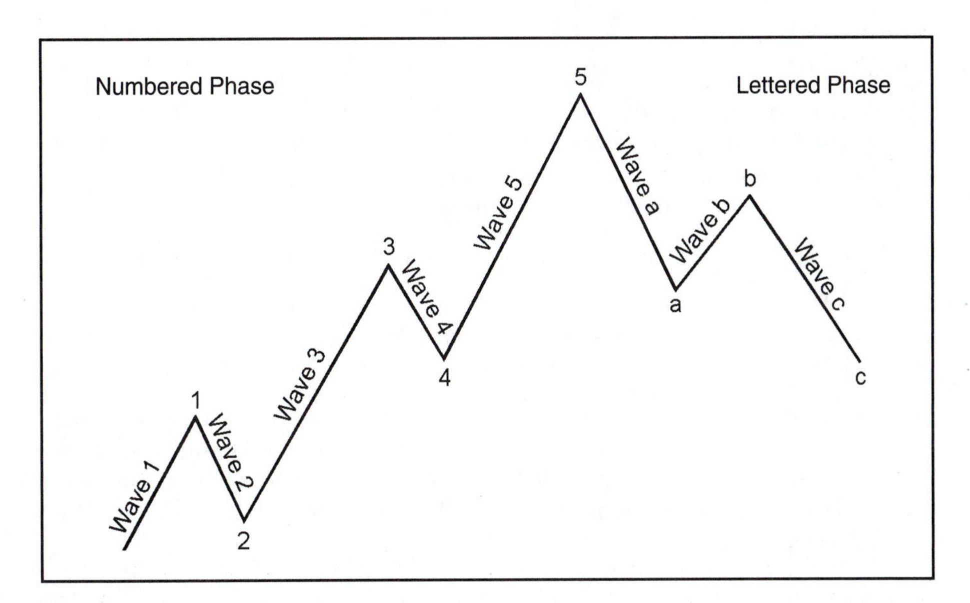

Figure 14.10 Elliott wave theory of five up waves and three down waves

New tools are constantly being developed. Both fractals and chaos theory are currently being used to try and observe patterns and waves. Because computers are becoming more powerful and personally affordable and extensive data bases are now available, new traders are entering the market. They are developing new tools and theories daily for price forecasting.

A Final Note

Price analysis is a frustrating activity. On any given day most analysts would agree that accurate price forecasting is impossible. Each analyst will tout the merits of his or her current tool or system and deride any others as useless or witchcraft. The most important question is, "Why bother?"

To effectively invest in the futures and options markets, either as a hedger or speculator, the trader does not need to do price analysis. However, a trader's effectiveness as an investor would improve with some type of price analysis for these reasons:

1. Price analysis forces the trader to stay current with where the market is and where it has been. The fear factor is reduced.
2. Traders who use price analysis generally have more trading and investing discipline, especially if a trading system is developed. The greed factor is reduced.
3. Watching and trying to understand prices helps an investor set stops and develop realistic pricing goals. The profit factor is increased.

PROBLEMS

1. Marie is a position trader. She currently believes that the price of gold will be bearish for a long time. She sells a futures contract on December gold for $360 per ounce and sets a stop loss order at $370. Of what use would past price charts of the December gold futures contract be to Marie?
2. How could Marie use both fundamental and technical price analysis with her position trades?
3. What is the relationship between options hedging and price analysis? between over-/underhedging with futures and price analysis?
4. Fundamental price analysis involves the use of economic factors to forecast price movements. Technical analysis ignores economic factors and simply says that where the market prices have been in the past is an indication of where they will be going in the future. They both seem very different. How are they the same?

APPENDIX A: Recommended Readings

Futures and Options

Arditti, Fred. *Derivatives.* Harvard Business School Press, Boston, MA, 1996.

Battley, Nick, Ed. *The Worlds of Futures and Options Markets.* Probus Publishing Company, Chicago, IL, 1993.

Bookstaber, Richard. *Option Pricing and Investment Strategies*, 3rd ed. Probus Publishing Company, Chicago, IL, 1991.

Colburn, James. *Trading in Options on Futures.* New York Institute of Finance, New York, NY, 1990.

Daigler, Robert. *Financial Futures and Options Markets.* HarperCollins, New York, NY, 1994.

Dodd, Mihel. *Trading Stock Index Options.* Probus Publishing Company, Chicago, IL, 1988.

Hull, John. *Options, Futures and Other Derivative Securities*, 2nd ed. Prentice Hall, Englewood Cliffs, NJ, 1993.

Klein, Robert, and Jess Lederman, Eds. *The Handbook of Derivatives and Synthetics.* Probus Publishing Company, Chicago, IL, 1994.

Labuszerwski, John, and John Nyhoff. *Trading Options on Futures.* John Wiley and Sons, Inc., New York, NY, 1988.

———. *Trading Financial Futures.* John Wiley and Sons, Inc., New York, NY, 1988.

Leuthold, Raymond, Joan Junkus, and Jean Cordier. *The Theory and Practices of Futures Markets.* Lexington Books, Lexington, MA, 1989.

Melamed, Leo. *Leo Melamed on the Markets.* John Wiley and Sons, Inc., New York, NY, 1993.

McCaffarty, Thomas, and Russell Wasendorf. *All About Futures.* Probus Publishing Company, Chicago, IL, 1992.

Purcell, Wayne. *Agricultural Futures and Options.* MacMillan Publishing Company, New York, NY, 1991.

Rotella, Robert. *The Elements of Successful Trading.* New York Institute of Finance, New York, NY, 1992.

Schwarz, Edward, Joanne Hill, and Thomas Schneeweis. *Financial Futures.* Irwin, Inc., Chicago, IL, 1986.

Tamarkin, Bob. *The Merc.* HarperBusiness, New York, NY, 1993.

Speculating

Greising, David, and Laurie Morse. *Brokers: Bagmen and Moles.* John Wiley and Sons, Inc., New York, NY, 1991.

Kleinfield, Sonny. *The Traders.* Holt, Rinehart and Winston, New York, NY, 1983.

Niederhoffer, Victor. *The Education of a Speculator.* John Wiley and Sons, Inc., New York, NY, 1997.

Schwager, Jack. *Market Wizards.* Harper and Row, New York, NY, 1990.

———. *The New Market Wizards.* HarperBusiness, New York, NY, 1992.

Price Analysis

New York Institute of Finance. *Technical Analysis: A Personal Seminar.* New York Financial Corporation, New York, NY, 1989.

Pistolese, Clifford. *Using Technical Analysis.* Probus Publishing Company, Chicago, IL, 1989.

Pring, Martin. *Technical Analysis Explained,* 3rd ed. McGraw-Hill, Inc., New York, NY, 1991.

Schwager, Jack. *On Futures, Fundamental Analysis.* John Wiley and Sons, Inc., New York, NY, 1995.

APPENDIX B: Recommended Internet Sites

All these sites have excellent information, plus many of them have great links to hundreds of other sites.

Exchanges

Chicago Mercantile Exchange, www.cme.com
Chicago Board of Trade, www.cbt.com
Kansas City Board of Trade, www.kcbt.com
MidAmerica Commodity Exchange, www.midam.com
Minneapolis Grain Exchange, www.mgex.com
New York Coffee, Sugar and Cocoa Exchange, www.csce.com
New York Cotton, Citrus, Finex and NYFE Exchange, www.nyce.com
New York Mercantile Exchange, www.nymex.com

Government

Commodity Futures Trading Commission, www.cftc.gov
U.S. Securities and Exchange Commission, www.sec.gov
Federal Reserve, www.bog.frb.fed.us

Organizations

National Futures Association, www.nfa.futures.org
International Swaps and Derivatives Association, www.isda.org

Other

First Internet National Directory, Commodity Futures Brokers Directory, www.findlists.com/futures/html/exchanges.html
Office of Futures and Options Research, University of Illinois, w3.aces.uiuc.edu/ACE/ofor/aboutofor.html
Waldermar's List, www.wsdinc.com/pgs_www/w7683.shtml
The Council (an Interactive Learning Mode for Futures), www.cpaee.com

APPENDIX C: Exchange Traded Futures and Options

(All listed futures contracts have options except where noted.)

Contract	Size	Delivery Months
Chicago Mercantile Exchange		
Agricultural		
Boneless beef	20,000 lbs.	Feb, Apr, Jun, Aug, Oct, Dec
Boneless beef trimmings	20,000 lbs.	Feb, Apr, Jun, Aug, Oct, Dec
Live cattle	40,000 lbs.	Feb, Apr, Jun, Aug, Oct, Dec
Feeder cattle	50,000 lbs.	Feb, Mar, Apr, May, Aug, Sep, Oct, Nov
Fluid milk	20,000 lbs.	All calendar months
Butter	40,000 lbs.	Feb, Mar, May, Jul, Sep, Oct
Lean hogs	40,000 lbs.	Feb, Apr, Jun, Jul, Aug, Oct, Dec
Pork bellies	40,000 lbs.	Jan, Mar, May, Jul, Aug, Sep, Nov
Random length lumber	80,000 bd. ft.	Jan, Mar, May, Jul, Sep, Nov
Oriented strand board	100,000 sq. ft.	Jan, Mar, May, Jul, Sep, Nov
Interest Rates		
Euro time deposit	$1,000,000	Mar, Jun, Sep, Dec
13-week Treasury bills	$1,000,000	Mar, Jun, Sep, Dec
1-year Treasury bills	$500,000	Mar, Jun, Sep, Dec

Contract	Size	Delivery Months
1-month LIBOR	$3,000,000	All calendar months
1-month Fed Fund rate	$3,000,000	All calendar months
Brady Bonds, Mexican Par	$1,000 × Price	Mar, Jun, Sep, Dec
Brady Bonds, Argentine FRB	$1,000 × Price	Mar, Jun, Sep, Dec
Brady Bonds, Brazilian C	$1,000 × Price	Mar, Jun, Sep, Dec
Brady Bonds, Brazilian EI	$1,000 × Price	Mar, Jun, Sep, Dec
3-month EuroMark	1,000,000DM	Mar, Jun, Sep, Dec
EuroYen	100,000,000Yen	Mar, Jun, Sep, Dec
91-day Mexican T-bills	2,000,000MxP	Mar, Jun, Sep, Dec
28-day Mexican interest rates	600,000MxP	All calendar months
Indices		
S & P 500	$500 × Index price	Mar, Jun, Sep, Dec
S & P MidCap 400	$400 × Index price	Mar, Jun, Sep, Dec
Nikkei	$5 × Nikkei 225 Price	Mar, Jun, Sep, Dec
Goldman Sachs Commodity Index	$250 × Index price	All calendar months
FT-SE 100 Share Index	$50 × Index price	Mar, Jun, Sep, Dec
Russell 8000 Stock Index	$500 × Index price	Mar, Jun, Sep, Dec
Major Market Index	$500 × Index price	Mar, Jun, Sep, Dec
S & P 500/BARRA Growth	$500 × Index price	Mar, Jun, Sep, Dec
S & P 500/BARRA Value	$500 × Index price	Mar, Jun, Sep, Dec
Nasdaq 100 Index	$100 × Index price	Mar, Jun, Sep, Dec
IPC Mexican Stock Index	$25 × Index price	Mar, Jun, Sep, Dec
Dow Jones Taiwan Stock Index	$250 × Index price	Mar, Jun, Sep, Dec
Currencies		
Australian Dollar	100,000 A$	Jan, Mar, Apr, Jun, Jul, Sep, Oct, Dec
Brazilian Real	100,000 Reals	All calendar months
British Pound	62,000 Pounds	Jan, Mar, Apr, Jun, Jul, Sep, Oct, Dec
Canadian Dollar	100,000 C$	Jan, Mar, Apr, Jun, Jul, Sep, Oct, Dec
Deutsche Mark	125,000 DM	Jan, Mar, Apr, Jun, Jul, Sep, Oct, Dec
French Franc	500,000 Francs	Jan, Mar, Apr, Jun, Jul, Sep, Oct, Dec

Contract	Size	Delivery Months
Japanese Yen	12,500,000 Yen	Jan, Mar, Apr, Jun, Jul, Sep, Oct, Dec
Mexican Peso	500,000 New Pesos	Mar, Jun, Sep, Dec
New Zealand Dollar	100,000 New NZ$	Mar, Jun, Sep, Dec
South African Rand	500,000 Rand	Mar, Jun, Sep, Dec
Swiss Franc	125,000 SFrancs	Mar, Jun, Sep, Dec

Chicago Board of Trade

Agriculture

Soybeans	5,000 bu.	Sep, Nov, Jan, Mar, May, Jul, Aug
Soybean oil	60,000 lbs.	Oct, Dec, Jan, Mar, May, Jul, Aug, Sep
Soybean meal	100 tons	Oct, Dec, Jan, Mar, May, Jul, Aug, Sep
Corn	5,000 bu.	Dec, Mar, May, Jul, Sept
Oat	5,000 bu.	Jul, Sep, Dec, Mar, May
Wheat	5,000 bu.	Jul, Sep, Dec, Mar, May
Rough rice	2,000 cwt.	Sep, Nov, Jan, Mar, May, Jul
Anhydrous ammonia	100 tons	Feb, Apr, Jun, Sep, Dec
Diammonium phosphate	100 tons	Mar, Jun, Sep, Dec

Metals

Gold	1 Gross Kilo	All calendar months
Gold	100 fine troy oz.	Feb, Apr, Jun, Aug, Oct, Dec
Silver	1,000 troy oz.	Feb, Apr, Jun, Aug, Oct, Dec
Silver	5,000 troy oz.	Feb, Apr, Jun, Aug, Oct, Dec

Contract	Size	Delivery Months
Financial		
U.S. Treasury bonds	$100,000 Par	Mar, Jun, Sep, Dec
U.S. Treasury notes, 10-year	$100,000 Par	Mar, Jun, Sep, Dec
U.S. Treasury notes, 5-year	$100,000 Par	Mar, Jun, Sep, Dec
U.S. Treasury notes, 2-year	$200,000 Par	Mar, Jun, Sep, Dec
Long-term municipal bonds	$1,000 × Price index	Mar, Jun, Sep, Dec
30-day Fed Funds	$5,000,000	All calendar months
PCS Catastrophe Insurance Options	Formula Based	Mar, Jun, Sep, Dec

Midamerica Commodity Exchange

Contract	Size	Delivery Months
Agricultural		
Corn	1,000 bu.	Mar, May, Jul, Sep, Dec
Oats	1,000 bu.	Mar, May, Jul, Sep, Dec
Wheat	1,000 bu.	Mar, May, Jul, Sep, Dec
Soybeans	1,000 bu.	Jan, Mar, May, Jul, Aug, Sep, Oct, Dec
Soybean meal	50 tons	Jan, Mar, May, Jul, Aug, Sep, Oct, Dec
Soybean oil	30,000 lbs.	Jan, Mar, May, Jul, Aug, Sep, Oct, Dec
Live cattle	20,000 lbs.	Feb, Apr, Jun, Aug, Oct, Dec
New York gold	33.2 fine troy oz.	All calendar months
New York silver	1,000 troy oz.	All calendar months
Platinum	25 troy oz.	All calendar months
Financial		
Australian Dollar	50,000 A$	Mar, Jun, Sep, Dec
British Pound	12,500 Pounds	Mar, Jun, Sep, Dec
Deutsche Marks	62,500 DM	Mar, Jun, Sep, Dec
Japanese Yen	6,250,000 Yen	Mar, Jun, Sep, Dec
Swiss Franc	62,500 SFrancs	Mar, Jun, Sep, Dec
Canadian Dollar	50,000 C$	Mar, Jun, Sep, Dec

Contract	Size	Delivery Months
Eurodollars	$500,000	Mar, Jun, Sep, Dec
U.S. Treasury notes, 10-year	$50,000 Par	Mar, Jun, Sep, Dec
U.S. Treasury notes, 5-year	$50,000 Par	Mar, Jun, Sep, Dec
U.S. Treasury bonds	$50,000 Par	Mar, Jun, Sep, Dec
U.S. Treasury bills	$500,000	Mar, Jun, Sep, Dec

Minneapolis Grain Exchange

Contract	Size	Delivery Months
Barley	180,000 lbs.	Mar, May, Jul, Sep, Dec
White wheat	5,000 bu.	Mar, May, Jul, Sep, Dec
Spring wheat	5,000 bu.	Mar, May, Jul, Sep, Dec
White shrimp	5,000 lbs.	All calendar months
Black Tiger shrimp	5,000 lbs.	All calendar months

Kansas City Board of Trade

Contract	Size	Delivery Months
Hard red winter wheat	5,000 bu.	Jul, Sep, Dec, Mar, May
Value Line Stock Index	$500 × Price index	Mar, Jun, Sep, Dec
Mini Value Line Stock Index	$100 × Price index	Mar, Jun, Sep, Dec
Western Natural Gas	10,000MMBtu	All calendar months

New York Coffee, Sugar and Cocoa Exchange

Contract	Size	Delivery Months
Coffee "C"	37,500 lbs.	Mar, May, Jul, Sep, Dec
Brazil Differential	37,500 lbs.	Mar, May, Jul, Sep, Dec
World Sugar No.11	37,500 lbs.	Mar, May, Jul, Oct
Domestic Sugar No.14	112,00 lbs.	Jan, Mar, May, Jul, Sep, Nov
World White Sugar	50 metric tons	Mar, May, Jul, Oct, Dec
Cocoa	10 metric tons	Mar, May, Jul, Sep, Dec

Contract	Size	Delivery Months
Milk	50,000 lbs.	Mar, May, Jul, Sep, Dec
Cheddar cheese	10,500 lbs.	Mar, May, Jul, Sep, Dec
Nonfat dry milk	11,000 lbs.	Mar, May, Jul, Sep, Dec
Butter	10,000 lbs.	Mar, May, Jul, Sep, Dec
BFP Milk	100,000 lbs.	Feb, Apr, Jun, Aug, Oct, Dec

New York Mercantile Exchange

Energy

Gulf Coast unleaded gasoline	42,000 gals.	All calendar months
New York Harbor unleaded gas	42,000 gals	All calendar months
Heating oil	42,000 gals.	All calendar months
Light sweet crude	42,000 gals	All calendar months
Sour crude	42,000 gals	All calendar months
Propane	42,000 gals.	All calendar months
Henry Hub natural gas	10,000 MMBtu	All calendar months
Permian Basin natural gas	10,000 MMBtu	All calendar months
Alberta natural gas	10,000 MMBtu	All calendar months
Crack Spreads Options	One put/One call	Mar, Jun, Sep, Dec
Electricity	736 mwh	All calendar months

Other

Euro Top 100 Index	$100 × Price index	Mar, Jun, Sep, Dec
Gold	100 fine troy oz.	All calendar months
Copper	25,000 lbs.	All calendar months
Silver	5,000 troy oz.	All calendar months
Palladium	100 troy oz.	All calendar months
Platinum	50 troy oz.	All calendar months

New York Cotton Exchange

Cotton	50,000 lbs.	Mar, May, Jul, Oct, Dec

Contract	Size	Delivery Months
Frozen concentrated orange juice	15,000 lbs.	Jan, Mar, May, Jul, Sep, Nov
Potatoes	80,000 lbs.	Jan, Mar, May, Jul, Sep, Nov
New York Composite Stock Index	$500 × Price index	Mar, Jun, Sep, Dec
Emerging Market Debt Index	$1000 × Price index	Mar, Jun, Sep, Dec
2YTN5YTR Treasury note	$500,000 Par 2-Yr. $250,000 Par 5-Yr.	Mar, Jun, Sep, Dec
CRB Bridge Futures Price Index	$500 × Price index	Jan, Feb, Apr, Jun, Aug, Nov
Cross Rate Futures		
Deutsche Marks/Japanese Yen	125,000 DM	Mar, Jun, Sep, Dec
Deutsche Marks/French Franc	125,000 DM	Mar, Jun, Sep, Dec
Deutsche Marks/Italian Lira	125,000 DM	Mar, Jun, Sep, Dec
Deutsche Marks/Swiss Franc	125,000 DM	Mar, Jun, Sep, Dec
Deutsche Marks/Spanish Peseta	125,000 DM	Mar, Jun, Sep, Dec
British Pound/Deutsche Mark	125,000 Pounds	Mar, Jun, Sep, Dec
British Pound/Japanese Yen	125,000 Pounds	Mar, Jun, Sep, Dec
British Pound/Swiss Franc	125,000 Pounds	Mar, Jun, Sep, Dec
U.S. Dollar Currency Paired		
U.S. Dollar/Deutsche Mark	$200,000	Mar, Jun, Sep, Dec
U.S. Dollar/Japanese Yen	$200,000	Mar, Jun, Sep, Dec
U.S. Dollar/Canadian Dollar	$200,000	Mar, Jun, Sep, Dec
U.S. Dollar/Swiss Franc	$200,000	Mar, Jun, Sep, Dec
British Pound/U.S. Dollar	125,000 Pounds	Mar, Jun, Sep, Dec
Australian Dollar/U.S. Dollar	200,000 A$	Mar, Jun, Sep, Dec
New Zealand Dollar/U.S. Dollar	200,000 NZ$	Mar, Jun, Sep, Dec
U.S. Dollar/S.African Rand	$100,000	Mar, Jun, Sep, Dec
U.S. Dollar/Singapore Dollar	250,000 S$	Mar, Jun, Sep, Dec
U.S. Dollar/Indonesian Rupiah	500,000,000 Rupiah	Mar, Jun, Sep, Dec
U.S. Dollar/Malaysian Ringgit	500,000 Ringgits	Mar, Jun, Sep, Dec
U.S. Dollar/Thai Baht	5,000,000 Bahts	Mar, Jun, Sep, Dec

Glossary

A

arbitrage—The process of simultaneously selling and buying in two or more markets to take advantage of a perceived difference.

at-the-money—A put or call option whose strike price is the same as the price of the underlying contract.

B

basis—The difference between a cash price and a futures price.

bear—A trader who believes the market price will fall.

brokers—Individuals who act on behalf of traders to buy or sell a futures or options contract.

bull—A trader who believes the market price will rise.

butterfly—A trade that simultaneously involves the purchase of two different futures delivery months and the selling of two of the same delivery months between the two purchases, as: Buy one May corn futures; sell two July corn futures; buy one September corn futures. The butterfly trade is used to take advantage of nearby prices strengthening relative to faraway prices. *See* **spreading**, **arbitrage**.

C

call—An option that the buyer has the right but not the obligation to purchase the underlying contract or commodity.

cash settlement—The process of discharging or offsetting a futures contract that has expired. The futures obligation is offset by calculating the difference between the final futures price and a final cash price (usually an average or index value).

chart analysis—*See* **technical analysis**.

clearinghouse—A third party between traders that assures the performance of each trader.

Commodity Futures Trading Commission (CFTC)—The federal regulatory agency for all futures and options contracts traded on organized exchanges. The CFTC was created in 1974 to replace the Commodity Exchange Authority. The CFTC is composed of five members appointed by the President and confirmed by the Senate.

cost of carry—The costs involved in storing from one time period to another. Usually is composed of actual storage costs, insurance, and the time value of money.

cracking—The process of converting crude oil into petroleum products.

cracking margin—The difference on a barrel basis between the price of a barrel of crude oil and the price of petroleum products.

cracking yield—The amount of petroleum products in a barrel of crude oil.

D

day trader—A person who holds an open position for no longer than a given trading day.

daily trading limit—The price limit, both high and low boundaries, that the futures price must trade within each day.

delivery—The process of offsetting an open futures position by going through the process of physical delivery or cash settlement.

delivery point—The designated place the commodity must be moved to in order to satisfy the terms of delivery.

double whammy—A hedge that is lifted before the cash position changes such that a loss on the futures contracts results, and then when the cash position is offset, a loss also occurs—a double loss.

E

econometrics—The use of statistics and mathematics to evaluate economic theories.

exercise—The conversion of an option into a demand for performance on the underlying contract.

F

floor broker—A holder of a seat on an exchange who trades in the pit for customers that issue orders.

floor trader—A holder of a seat on an exchange who trades in the pit for his own account.

forward sell—The act of selling an item for future delivery.

futures commission merchant (FCM)—A person who buys/sells futures contracts for a client for a fee.

futures contract—A legal obligation to deliver (a sell) or accept delivery (a buy) of a specific commodity with contract terms standardized.

H

hedge—The process of shifting price risks in the cash market to the futures market by simultaneously holding opposite positions in the cash and futures markets.

I

in-the-money—A put (call) option that has a strike price that is higher (lower) than the price of the underlying contract.

intrinsic value—The numeric difference between an in-the-money strike price and the price of the underlying contract.

inverted market—Distant futures delivery month prices are lower than nearby month's price.

L

last trading day—The last day that a futures or options contract can be traded before the contract month expires.

long—An initial buy position of a futures or options contract or the physical ownership of the cash commodity.

long hedge—An initial purchase of a futures contract or an initial purchase of a call option contract used to protect against an initial forward sell in the cash market.

M

maintenance margin—The predetermined amount of the initial margin that triggers a margin call signifying that the position has lost enough money to require more cash to hold the position.

margin—The initial amount of good faith cash that must be posted with a broker to enter into a futures or options position.

margin call—The amount of money that must be deposited with the broker to maintain a losing futures or options position.

marking to market—The process of updating the value of the contract each day to reflect the change in market price.

N

National Futures Association (NFA)—Formed by Congress in 1982 to regulate the futures and options markets via self-regulation with dues from members.

nearby—The closest futures delivery month of the current date.

net hedge price—The net price a hedger receives considering both cash and futures transactions.

O

offset—The opposite action taken to get out of an initial futures or options position.

open interest—The number of open (not yet offset) contracts.

open outcry—The process of obtaining an established price for a futures or options contract in the trading pit. It must be both by voice and hand signals.

option—A contract that gives the buyer the right but not the obligation to obtain an item/service. The seller of the contract has an obligation to perform, should the buyer exercise the right.

out-of-the-money—A put (call) option that has a strike price that is less (more) than the price of the underlying contract.

P

paper gain—The amount that the current futures price is different than the initial price such that a gain would occur if offset.

paper loss—The amount that the current futures price is different than the initial price such that a loss would occur if offset.

pit—The actual place where open outcry occurs and the futures and options contracts trade at the exchange.

position limit—The maximum number of contracts that a trader can own and/or control of each commodity.

premium—The price that an option contract trades for a given strike price.

put—An option contract that the buyer has the right but not the obligation to have a sell position on the underlying contract.

R

reverse spread—A simultaneous position that involves having a sell position on the nearby futures contract and a buy position in a more-distant futures contract.

rolling hedge—The process of offsetting the futures position and replacing with a more-distant futures position because the cash position has not been altered.

roundturn—*See* **offset**.

S

scalper—A trader in the pit who holds a position only briefly and trades on small price moves.

seat—A membership at an exchange that gives the owner or leasor the right to trade at that exchange.

settle price—The final price used each day to value futures and options contracts. It involves the actual last traded price and a weighted average of the last trades during the last few minutes of the trading day.

short—An initial sell position with a futures or options contract or the act of forward selling a cash position.

short hedge—An initial sell in the futures market to offset a long cash position.

speculator—Someone who trades futures or options contracts with the intention of making a profit and does not own and/or control the actual cash commodity.

spot market—The actual cash market at the current moment.

spread—The simultaneous position involving an initial buy of the nearby futures contract and an initial sell of a more distant month.

stop—The placing of an order to offset an initial futures position if a certain price level is reached.

strike price—The price that an option contract will be converted into the underlying position.

T

target price—The estimate of a net hedge price.

technical analysis—The belief that futures price direction can be determined by past price movements.

tick—The least value change that can occur in the price movement of a futures or options contract.

time value—The amount of the option premium that reflects the trader's expectations of future value. The net difference between the intrinsic value and the option premium for in-the-money options. Time value and the option premium are the same for at-the-money options and out-of-the money options.

time value of money—The opportunity cost of money.

to-bid price—A cash offered price derived from a futures price.

U

underlying contract—With options on futures contracts, it refers to the futures contract. With options on the actual commodity, it refers to the actual commodity.

V

volume—The number of total contracts traded.

W

whip sawed—The process of entering a futures or options position and then offsetting with a small loss or gain without a clear trend in price. Occurs in flat price movement markets.

Index

NOTE: Page references in **boldface** refer to figures/tables.

I

K

L

M

N

O